# BASIC CARPENTRY TECHNIQUES

Created and designed by the
editorial staff of ORTHO Books

Written by
T. Jeff Williams

Designed by
Craig Bergquist

Front Cover Photograph by
Fred Lyon

Illustrations by
Ron Hildebrand

**Ortho Books**

Publisher
Robert L. Iacopi

Editorial Director
Min S. Yee

Managing Editors
Anne Coolman
Michael D. Smith

System Manager
Mark Zielinski

Senior Editor
Sally W. Smith

Editors
Jim Beley
Diane Snow
Deni Stein

System Assistant
William F. Yusavage

Production Manager
Laurie Sheldon

Photographers
Laurie A. Black
Michael D. McKinley

Photo Editors
Anne Dickson-Pederson
Pam Peirce

Production Editor
Alice E. Mace

Production Assistant
Darcie S. Furlan

National Sales Manager
Garry P. Wellman

Operations/Distribution
William T. Pletcher

Operations Assistant
Donna M. White

Administrative Assistant
Georgiann Wright

Address all inquiries to:
Ortho Books
Chevron Chemical Company
Consumer Products Division
575 Market Street
San Francisco, CA 94105



Printed in August, 1981

2 3 4 5 6 7 8 9
84 85 86 87 88 89

ISBN 0-917102-95-9

Library of Congress Catalog Card
Number 80-85220

**Chevron Chemical Company**
575 Market Street, San Francisco, CA 94105

**Graphic Production by**

CABA Design Office
San Francisco, CA

**Typography by**

Vera Allen Composition
Castro Valley, CA

**Illustration Assistants**

Rhonda Hildebrand
Shandis McKray

**Copyediting by**

Editcetera
Berkeley, CA



**Consultants**

Frank W. Benda
John H. Palmer
San Francisco, CA

Robert A. McMahon
West Sacramento, CA

Euel Keith Construction
Roseville, CA

**Photographs**

Fred Kaplan
Pages 14, 64, and 74
Back Cover, upper left

Josephine Coatsworth
Page 1
Back cover, lower left

Michael Landis
Pages 4, 5, 15, 65, and 75
Back Cover, upper and lower right

Robert Wesley
Wesley Woodworking

**Front Cover photograph:**
A toolbox as small as this one can hold all the tools you need to build an entire house. Detailed information on all basic carpentry tools begins on page 15.

**Page 1 photograph:**
Writer T. Jeff Williams at work on one of his many construction projects.

**Back Cover Photographs:**
*Upper left*—Cutting lumber at the sawmill. See pages 6–7 to learn how trees become lumber.

*Upper right*—Framing a wall for the playhouse. See page 90.

*Lower left*—Cutting the birdsmouth in a rafter. See page 97.

*Lower right*—Completed playhouse. See pages 78–79 for a cut-away view.

# BASIC CARPENTRY TECHNIQUES

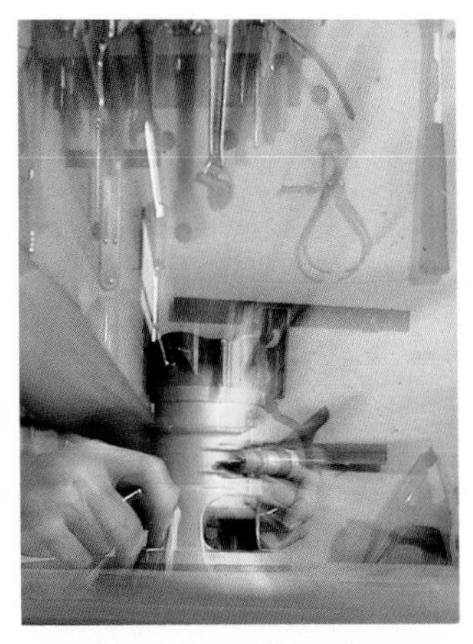

# FIRST STEPS IN CARPENTRY

One of the first steps in carpentry is figuring out what lumber to buy. Illustrated descriptions of how lumber is cut and graded along with charts of wood species and characteristics will help you make the right choices on all your building projects.

## Getting Started

There are basically two types of carpentry: construction and cabinetry. *Construction carpentry* generally refers to using unfinished woods in medium- to large-scale projects. This type of carpentry is further sub-divided into two types: rough and finish. *Rough* carpentry includes foundations, wall, floor and roof framing, installation of windows and exterior doors, and some exterior finish work. *Finish* carpentry includes interior floor, ceiling and wall coverings, trimwork, building stairs and installation of cabinets and interior doors. *Cabinetry* is associated with painstaking craftsmanship on a product with a highly finished appearance, complete with sealants, sandings, and stains. It includes mill work, such as shaping moldings and trims. Since tools for both types of carpentry are generally the same, a large portion of this book is devoted to describing these tools and to illustrating the techniques for using them. Special attention is given to sawing—from the simplest crosscut to the special cuts required for making joints. No matter what kind of carpentry you do, you'll want to know what tools to use and how to use them with skill.

Beyond the skillful use of tools lies the actual process of construction—putting all the pieces together. This book concentrates primarily on the construction process for rough carpentry and includes some finish carpentry techniques. It demonstrates the step-by-step process of building a good workshop, including workbench and drawer construction, and moves on to actual house construction techniques. Starting with the foundation of a building, it continues right through to roofing and installing the front door. (For construction techniques used in cabinetry, see Ortho's book *Wood Projects for the Home*.)

If you have wanted to build a separate office or studio, or a large playhouse like the one described in Chapter 4, this book will enable you to do it. The basic techniques used in constructing the playhouse are the same as those used in building a full-scale house, so all you need to get started is confidence, willingness, a piece of property, and this book.

However, if you are a novice and your first project is complex, you may also want to consider working with an experienced carpenter. You will probably complete your project with more efficiency, and the time you save may offset any expenses incurred by hiring someone. Further, working side by side with an experienced carpenter will give you more building knowledge than you can gain by working alone. Experience is an excellent teacher, but experience under a good teacher is even better.

## Buying Tools

If you are just starting as a builder, don't despair at the variety of tools. Start out with a few of the most essential: a saw, hammer, drill, measuring tape, and some screwdrivers. Add more tools as you really need them. Even if

◀

**It is your vision of a completed project that leads you to your first steps in carpentry.**

**Framing the playhouse shown in its finished form at left. Construction details start on page 78.**

# LUMBER QUALITY

you are only planning to do small weekend carpentry projects, buy good tools. Get the best you can afford—or, better yet, get tools that cost a little *more* than you think you can afford. This will save you the cost of a replacement when the cheap tool breaks. Good quality always saves you money in the long run.

Good tools are also essential to good workmanship, and they contribute to your peace of mind. If you have a cheap handsaw that barely cuts and keeps binding, you will quickly tire of your project. Carpentry should be a labor of love, and good tools will help you maintain that level of enjoyment.

## Building Permits

Building permits are normally required whenever you make any structural changes or additions on your house, or start a new building, such as the playhouse/studio described in this book. Changes within the house that do not alter the building's structure, such as cabinets, partitions, or dropped ceilings, do not generally require a permit. But if you want to put in a new and larger window that involves altering the wall of your house, you will need a building permit. Building permits assure that your construction will be safe and sound, and enable your local government to reassess your property taxes according to the value of the improvements you make.

Building codes usually cover all aspects of construction and structural alteration, including material specifications, the details of foundations and framing, and design factors such as room size, lighting, and ventilation. Codes vary widely across the country. However, the remainder of this book gives you the general information necessary to understand your local codes.

### Submitting Your Construction Plans

To secure a permit, you must submit a set of plans to the building inspector's office. Call before you go to find out how many copies of your plans they need. Plans do not always have to be professionally drawn; they just need to be complete in all details. If your project is fairly small, such as a playhouse or room addition, you can probably draw up the plans yourself. (See pages 76–77.)

Before you begin your plans, talk with some experienced builders or carpenters in your neighborhood. They can quickly tell you what the local codes require. Then go to the inspector's office and ask to read the codes on residential construction, checking the information you have gathered from friends.

Remember that the building inspector's office administers codes regulating electrical wiring, water pipes, and the sewage system as well as construction materials and techniques. While these codes may seem complicated at first, you will soon become familiar with them, and the information they contain can help you avoid costly errors or oversights.

When your plans are ready, take them to the inspector's office. Any deficiencies will be quickly spotted, and you may be sent back to the drawing board. Don't get upset. Just consider your first meeting a good lesson in construction, and keep this attitude throughout your dealings with the inspector's office. The idea is to get as much help and advice as possible.

**Plain Sawed Lumber**

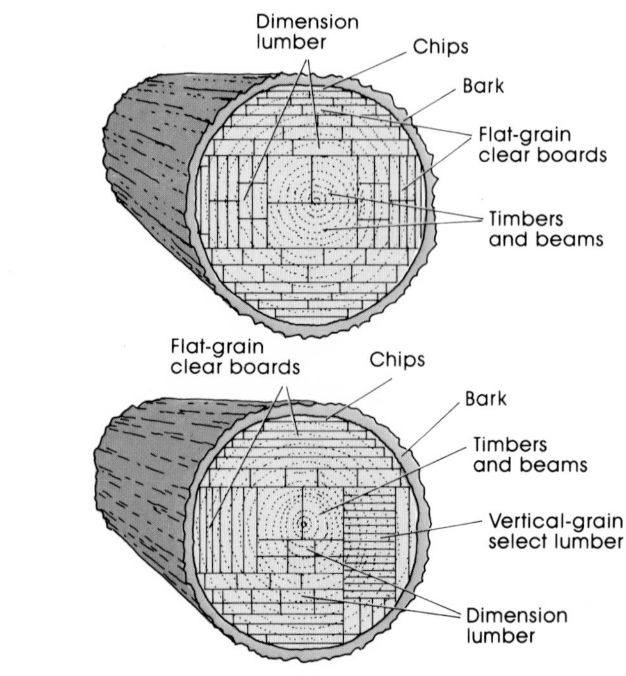


**Quarter Sawed Lumber**

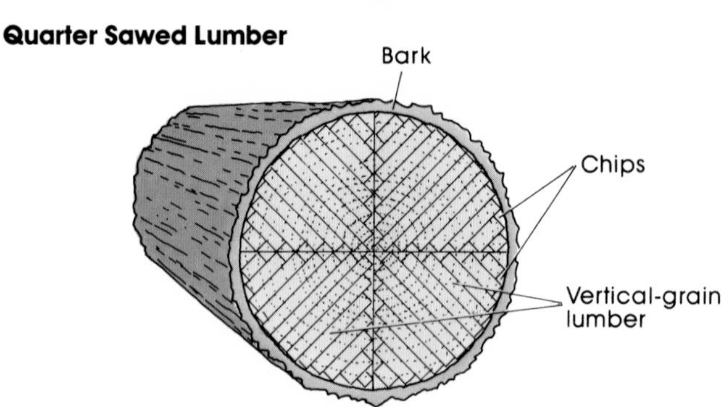


After debarking, a log is cut in either of two basic ways. The most common is called flat-grained, slash-, or plain-sawed. This process results in less waste and therefore the wood is less expensive. The other way, called edge-grained or quarter-sawed, results in lumber that is less likely to warp and is more beautifully grained in most species.

There are further variations to these basic cutting methods, depending on the species, size of log, and the kind of lumber ordered.

---

The quality of lumber is important to your project. It is affected by three major factors: the species of tree; the part of the tree from which the lumber is cut; and how the lumber is dried, surfaced, and stored. All of these factors should be considered when you select lumber. (See chart on pages 12–13.)

## From Trees to Lumber

Lumber is divided into two major categories: hardwoods and softwoods. Hardwood comes from trees that lose their leaves in the winter and is primarily used where a combination of strength and beauty is important, such as in furniture and floors. It is rarely used in construction carpentry but is discussed in Ortho's book *Wood Projects for the Home*.

Softwood comes from evergreens and provides virtually all the construction lumber in the United States and Canada. Softwood species vary greatly. A pine is not simply a pine. For example, the southern yellow pine is markedly stronger than the ponderosa pine that grows in Oregon and California. So you need to consider the relative strengths of different species before selecting the lumber for load-bearing structures.

Lumber from different species also varies in the way it takes stains. For example, pines take stains and paints better than firs—an important consideration if you are choosing molding or house siding.

## How Lumber is Cut

Generally, the first cut of the saw takes a slab off the top. This is mostly *sapwood*, the outer, fast-growing part of the tree that is full of pitch. Because it is poor material for lumber, it is normally turned into chips to make pulp and paper.

After one or two more 1-inch-thick cuts, the remaining cuts are determined. As a tree grows, lower limbs are increasingly shaded and eventually fall off, creating a knot. New wood continually covers the knot, burying it deeper in the tree. The deeper into the tree, the bigger the knots. The wood toward the center is cut into larger and larger beams, such as 4 by 4s and up, to overcome the inherent weakness of the knots.

A typical log produces the following general lumber cuts:

**Dimension lumber** is 2 to 4 inches thick and 2 or more inches wide. It is used in all types of construction and is graded for strength rather than appearance. (See discussion of grading on page 8.)

**Boards,** which are 1 inch thick, come in all widths up to 12 inches. They are generally used for cabinets, shelves, crates, or exterior siding. Boards are normally graded for appearance rather than strength.

**Timbers and beams** are 5 by 5 inches or larger and are mostly used for supporting framework. While generally graded for strength, they can also be graded for appearance, to be used in open beam ceilings, for example.

**Select and finish** lumber is usually found near the outer part of the tree and is largely free of knots and blemishes. This is top-quality lumber and is used for cabinets or wherever appearance is important.

## Drying and Surfacing Lumber

**"Green" lumber,** from a freshly felled tree, contains a high percentage of moisture. It makes the lumber heavy, difficult to work with, and subject to rot and twisting unless properly dried. Lumber is normally dried by air or in a kiln at the mill site before being shipped to retail outlets. Sometimes, however, it is shipped to the lumberyards while still green. The lumberyards are then supposed to stack it properly and let it air dry. If they don't do a good job, the lumber will warp, so you should select your own lumber whenever possible.

**Air-dried lumber** is stacked in sheds with strips of wood between each layer to ensure good air circulation. If your lumber is going to be sitting around for a week or more, you should stack it in the same way to let it dry further.

**Kiln-dried lumber** is placed in kilns, or ovens, to dry it more thoroughly than air drying can. This gives you better lumber, but at a higher cost. Generally, only high-quality wood to be used for cabinets or paneling is kiln dried.

As wood dries, it shrinks. The way the board was cut from the log will affect this shrinkage. Look at the end of a board. If the grain is parallel or nearly parallel to the face, it is a *flat grain* board and will shrink the most.

If the grain runs at right angles to the face of the board, it is known as *vertical grain*. This has less shrinkage than flat grain and is somewhat stronger and more expensive.

As the lumber dries, it will also begin to *twist, crook, bow,* or *cup* unless carefully stacked.

Lumber is normally stacked and dried while it is still *rough cut*, which means just as it was cut from the log. After drying, it is then *dressed*, or *surfaced*, to smooth it out and make all boards of standard dimensions. Thus, a 2-by-4 rough board measures the full, or *nominal*, 2 by 4 inches. Surfacing reduces it to a standard size, called *actual*, which is 1½ by 3½ inches. Lumber is always referred to by its nominal size.

### Nominal Lumber Sizes and Their Actual Equivalents

| Nominal | Actual |
|---|---|
| 1 × 2 | ¾ × 1½ |
| 1 × 3 | ¾ × 2½ |
| 1 × 4 | ¾ × 3½ |
| 1 × 6 | ¾ × 5½ |
| 1 × 8 | ¾ × 7¼ |
| 1 × 10 | ¾ × 9¼ |
| 1 × 12 | ¾ × 11¼ |
| 2 × 2 | 1½ × 1½ |
| 2 × 3 | 1½ × 2½ |
| 2 × 4 | 1½ × 3½ |
| 2 × 6 | 1½ × 5½ |
| 2 × 8 | 1½ × 7¼ |
| 2 × 10 | 1½ × 9¼ |
| 2 × 12 | 1½ × 11¼ |
| 4 × 4 | 3½ × 3½ |

After the lumber has been dried and surfaced, its moisture content is checked. If the wood has a moisture content of 20 percent or more, the lumber is stamped S-GRN (surfaced, green lumber). If it is 19 percent or less, it is stamped S-DRY (surfaced, dry lumber). If the lumber contains 15 percent or less moisture, usually obtained by kiln drying, it is stamped MC-15 (moisture content 15 percent or less). Don't buy green lumber unless it is markedly cheaper and you are prepared to air dry it yourself.

Most lumber you buy at a retail outlet will be surfaced on all four sides, though some is not. The fewer sides surfaced, the cheaper it will be. The one notable example is redwood, which can commonly be found rough (unsurfaced) or just surfaced on both faces.This is somewhat cheaper, and if you don't mind the rough edges, it is perfectly satisfactory.

**Siding**

Surfaced lumber, particularly redwood, is stamped according to how many sides are surfaced with one of the following marks: S1S (surfaced 1 side), S2S, S3S, or S4S.

**Worked lumber** has been milled for specific purposes. Three common types of worked lumber are illustrated above.

## Grading Categories

Lumber is graded by several independent agencies in the West, the North Central States, the South, and the East. They follow national grading rules established by the U.S. Department of Commerce. The purpose of grading is to give buyers a reliable yardstick for determining the quality of lumber. Strength and appearance are the basic criteria for grading.

Most lumber carries a stamp that tells the buyer which firm graded it, the mill that produced it, the quality of that board, how dry it is, and the wood species. When you select wood for a project, the two most important aspects of the lumber grade stamp to consider are the grade and the species.

There are two major grades of softwood lumber: select lumber, which is primarily graded for appearance rather than strength, and common lumber, which is primarily graded for structural strength. Select lumber is normally used for paneling or cabinetry, while common lumber is used for construction.

## Select Lumber

Select lumber is graded with letters—A, B, C, D, and E.

**Grade A** is clear lumber; that is, it has no knots or blemishes. It is expensive and often difficult to find.

**Grade B** has some small knots and slight blemishes. Like Grade A, it is usually given a clear finish to show off the fine grain. Some Grade A is often found in the same bin with Grade B, which is then designated "Grade B and Better."

**Grade C** has slightly larger knots and more blemishes than grade B but is still a good quality lumber for cabinetwork.

**Grades D and E** contain increasingly larger knots and knotholes and would be used largely as backing on a cabinet or in other areas where they would not be highly visible.

Your decision in buying select lumber will be influenced by how fine you want the wood to be and what you are willing to pay for it.

**Lumber Grading Stamp**

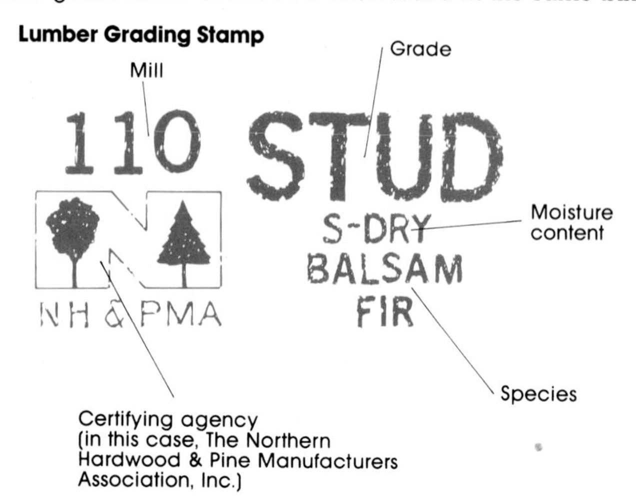


## Common Lumber

Common lumber is broken down into four different categories: (1) select structural, (2) structural joists and planks, (3) light framing, and (4) stud grade.

**1. Select structural lumber** is the highest grade in terms of both appearance and strength. However, in all these categories, strength is the first consideration in the grading.

Select structural lumber is graded by numbers that are included in the grading stamp. *Number 1 grade* is recommended where good appearance is required along with strength and stiffness. *Number 2 grade* has a less pleasing appearance but still retains high strength. It is recommended for most construction uses. *Number 3 grade* is recommended for general construction purposes that do not require the high strength necessary for floor joists and rafters.

**2. Structural joists and planks** are also graded with numbers, according to the same guidelines used for select structural lumber.

**3. Light framing lumber,** which is the most widely used category of lumber for framing houses, is broken into three different grades. *Construction grade* is the top of the line here: strong, stiff lumber with relatively few knots. *Standard grade* compares favorably with construction grade but has more knots and somewhat less strength. However, it is also cheaper than construction grade. *Most building codes require that lumber used in houses be standard or better.*

*Utility grade* is the last category of light framing lumber. This is generally quite knotty and weak. It can be used for temporary bracing or for blocking between studs and joists, but should not be used to bear loads, such as in the side walls of houses.

**4. Stud grade lumber** is specially selected for use as studs in house framing. It is generally quite straight, which helps the builder put up a straight wall. Stud grade lumber is limited to lengths of 10 feet or less.

## Species Identification

In the grading stamp, two or more species are often listed. This is because trees with similar characteristics often grow side by side. Rather than trying to sort out every tree, the mills simply lump together wood that is very similar. Thus on the West Coast you will often see lumber stamped "Doug Fir-L," which means it is either Douglas fir or larch.

When you select lumber for use where it will not be visible, such as in framing a house, species is less important than the grade of lumber. It only has to be strong enough to do the job. However, species becomes a more important consideration where the wood is highly visible. You will want to consider the beauty of the grain, workability, and painting or staining characteristics. As a rule of thumb in this matter, pine is a good choice for trim work because it is easily worked and readily takes paint. (See box on page 11).

## Plywood

Plywood is made by rotating 8½-foot-long logs at high speeds against a steel knife edge of the same length. The log is peeled in a continuous sheet of thin veneer, much like unwinding a roll of paper towels. This veneer is then dried and cut into plies, or sheets, and glued together with the plies at right angles to give plywood its great strength. It is always put together with an odd number of plies so that the grain on the front and back runs in the same direction.

Plywood generally comes in thicknesses of ¼, ⅜, ½, ⅝, ¾, and 1 inch. Standard sheets of plywood are 4 by 8 feet, but longer and thicker panels can be ordered.

Plywood is usually stamped as either exterior grade or interior grade. *Exterior grade* plywood is made with a waterproof glue, and its outer veneers are made from more water-resistant wood. *Interior grade* plywood is also usually made with waterproof glue, but its veneers will not weather as well.

Both softwoods and hardwoods are used in making plywood, but hardwood plywood is used only for finish work such as cabinets or paneling and is considered in Ortho's book *Wood Projects for the Home*.

### Plywood Grades

Softwood plywood comes in a confusing array of grades. But each sheet carries a grading stamp, which tells you all about that piece, if you know how to read it.

Plywood is broken into two basic categories: *appearance grades,* such as for siding or paneling, and *engineered grades,* where structural qualities are important, such as in concrete forms, sheathing, or subfloors.

Typically, the grade stamp includes two large letters separated by a dash, such as C-D. The first letter designates the quality of the front ply, and the second designates the quality of the back ply. These letters, in order of descending quality, are N, A, B, C, D. The finest appearances are N and A. N is free of any defects, A has a few, and B has more. C grade is the minimum grade allowed on exterior plywood. D grade is used only for the backs or inner plies of interior plywood.

For engineered plywood, the stamp also includes two numbers separated by a slash if the plywood is to be used for siding or flooring. These numbers are span indexes for either rafters or floor joists. The first number refers to permissible rafter spacing in inches, and the second to permissible joist spacing. A number such as 24/16 means you could use that sheet of plywood as roof sheathing if the rafters were no more than 24 inches apart, or as subflooring if the joists were no more than 16 inches apart.

Plywood has varying degrees of strength and stiffness.

**Plywood Grading Stamp**

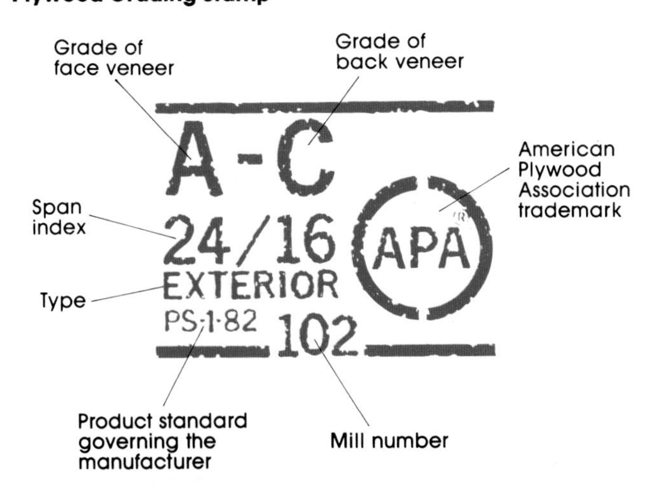


**Plywood Veneer Grades**

This chart with its description of the six different levels of veneer grades on plywood, listed in descending order of quality, was made available by the American Plywood Association.

|  |  |
|---|---|
| **N** | Smooth surface "natural finish" veneer. Made of select grade wood that is either all heartwood or all sapwood. Free of open defects. No more than six repairs permitted in each 4 × 8 panel; each must be made of wood, must be parallel to the grain, and must be matched to the grain and color of the panel. This is top quality grade. |
| **A** | Smooth and paintable with no more than eighteen neatly made repairs; each must be made parallel to the grain. Can be used for natural finish in applications that are not too demanding. |
| **B** | Solid surface with circular repair plugs, shims, and tight knots up to 1 inch across the grain allowed. Some minor splits permitted. |
| **C plugged** | Improved C veneer. Splits limited to ⅛-inch width with knotholes and borer (insect) holes limited to ¼ – ½ inch. Some synthetic repair and broken grain permitted. |
| **C** | Tight knots to 1½ inch allowed. Knotholes permitted up to 1 inch across the grain and some to 1½ inch if the total width of the knots and knotholes is within specific limits. Permissible to have synthetic or wood repairs, discoloration, and sanding defects if they do not impair the strength. Limited splits are allowed. Stitching — the process of sewing random-sized pieces of plywood together — is permitted. |
| **D** | Knots and knotholes across the grain and up to 2½ inches wide are allowed. Within specified limits they can be up to 3 inches wide. Stitching is permitted as is a limited number of splits. This level is limited to interior grades of plywood. |

Several types of wood may go into a sheet of plywood, and these woods are selected from more than 70 different species. For engineered plywood you can check the *group number* to determine the strength. Group numbers range from 1 to 5, with 1 being the strongest. Appearance plywood does not usually carry these group numbers.

There is also a separate category of plywood referred to as *shop grade*. Most lumberyards carry it, and it is about the cheapest plywood you can buy. Shop grade can carry any grade stamp but is rejected in the final checking because of imperfections, such as glue failure or split veneer. This grade will save you money, particularly where appearance is not important. Select your own to get the best of what is available.

## Hardboard

Hardboard is generally cheaper than plywood and is made from waste wood material. Lignin, a substance that holds wood fibers together, is removed from wood in paper production. The lignin and wood fibers are compressed under tremendous pressure to form the hardboard. It normally comes in 4 by 8-foot sheets, with one side smooth and the other waffled. It is most commonly used for drawer bottoms, cabinet backing, or covering rough plywood floors prior to laying carpeting. When manufactured with a series of holes 1 inch apart, it is called pegboard and is used in work centers for hanging tools.

Since hardboard contains abrasive materials such as glue, it will quickly dull a saw blade. Use carbide-tipped blades on your power saw to offset this problem.

### Choosing Wood Species

Wood species commonly recommended for the various elements of house construction are listed below. The lumber you choose will be determined in part by the region in which you live and what is available there. This list was prepared by the Western Wood Products Association in Portland, Oregon, the largest lumber-grading association in the country.

**FRAMING** (joists, beams, studs): Douglas fir, white fir, and other western firs, western and eastern hemlock, southern pine, larch, and spruce

**SIDING:** Western red cedar, northern white cedar, Douglas fir, redwood, southern cypress

**PANELING:** All pines, western red cedar, redwood, Douglas fir, western hemlock, larch, southern cypress

**CEILING DECKING:** White pine, lodgepole pine, Douglas fir

**WOODWORK** (windows, shutters, molding): Ponderosa pine (the most versatile pine species), sugar pine, southern pine, Douglas fir, white fir, redwood, larch, oak, maple, gum, walnut

**DOORS:** Douglas fir, western hemlock, pine, maple, birch

**FLOORING:** Maple, oak, beech, gum, Douglas fir, western hemlock

**STAIR TREADS:** Douglas fir (vertical grain for more strength), oak

**SHELVING:** All pines, Douglas fir, western hemlock

**CABINETWORK:** Douglas fir, western hemlock, cedar, pine, hackberry, walnut, oak

**DECKS, RAILINGS:** Redwood, western red cedar, southern cypress, Douglas fir, larch, lodgepole pine (All these need to be treated with a preservative when in contact with the ground.)

## Particleboard

Particleboard is another useful product made from sawdust and planer shavings, material that mills used to burn. Like hardboard, the 4-by-8-foot sheet is formed under great pressure. It is much weaker than hardboard, but it is cheaper than either hardboard or plywood. It is widely used for cabinet backing and shelves of lower quality, and it can also be laid over rough plywood subflooring before carpeting.

Particleboard does not hold nails well, but this can be offset by using glue whenever possible. Do not let it remain wet for any length of time or it will begin to disintegrate, unless you specifically order a water-resistant variety.

## Selecting Your Own Lumber

Lumber is usually stacked in large covered bins according to species, size, length, and grade. When you find the bin you want, slide each piece out and give it a quick but careful look. Using the problem boards illustrated here as a guide on what to avoid, first turn the board on edge and sight along the top to see if it is bowed, twisted, or crooked. Reject such boards when possible, although your choice may be limited. If the lumber will not be visible in your

**Problem Boards**

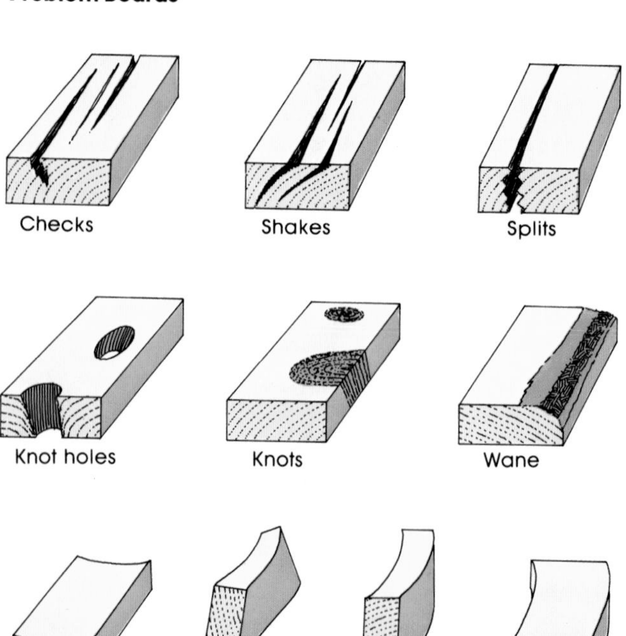


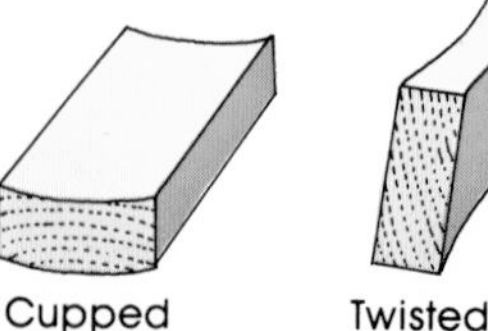
Cupped

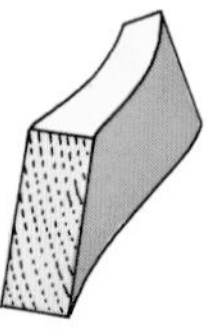
Twisted

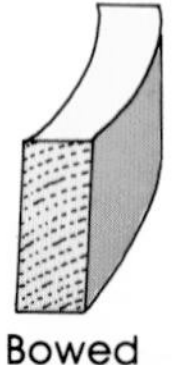
Bowed

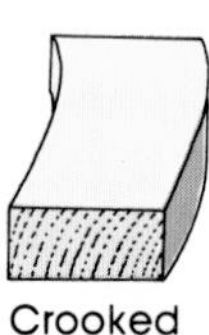
Crooked

construction project, don't worry about wanes, since they will not weaken a board much. Look for any loose knots or splits. Avoid boards that are loaded with pitch (resin produced by certain trees) since they are heavy and tend to split as they dry.

When possible, avoid green wood. Make sure that the board is stamped S-DRY, which means it has 19 percent or less moisture. A stamp saying MC-15 means it has 15 percent or less moisture. If you don't like what you see, talk to the counterman and find out whether a new shipment will be coming in soon so you will have a better selection. You should also shop at several lumberyards to compare prices and quality.

When you undertake a large project, such as a house, it is too time-consuming for you to select every piece of lumber. Instead, place your order and let the yard employees gather it up and deliver it. You can increase your chances of getting good quality lumber by explaining your needs carefully to the counterman. Ask if you can return unusable pieces. And try to become acquainted with the yard people who will be putting your order together. They can help.

## Estimating Lumber Needs

Always buy the smallest quantity possible to avoid waste and the lowest grade permitted to save money. If you have a fairly simple project, such as a shed, playhouse, or small barn, the best way to estimate your needs is simply to count up all the pieces from your plans. If you have professional architectural drawings for larger projects, a lumber list is generally included. If it isn't, ask the manager at the yard where you will buy all your lumber to work out the list for you. Since he is going to make a big sale, he may be willing to help you with it.

### Board Feet

Lumber is sold either by the *lineal foot* or by the *board foot*. Commonly, the shorter lengths of finish wood are sold by the lineal or running foot (regardless of size), while lumber sold in volume is based on the board foot. A board foot is 1 inch thick, 12 inches wide, and 12 inches long. If you cut that board in half and sandwich it into a 2 by 6 that is 12 inches long, you still have a board foot.

To calculate board feet, use this formula:

$$\frac{\text{Thickness (in.)} \times \text{Width (in.)} \times \text{Length (ft.)}}{12} = \text{Board feet}$$

or

$$\frac{T'' \times W'' \times L'}{12} = \text{Board feet}$$

Example: How many board feet are in an 8-foot-long 2 by 4 (written at the lumberyard as 2 × 4 × 8)?

$$\text{Answer: } 2 \times 4 \times 8 = \frac{64}{12} = 5.33$$

To calculate board feet, always use the nominal rather than the actual size of the lumber.

Lumber prices are calculated per thousand board feet. When you are ready to purchase a large amount of lumber, call several yards and compare prices by asking what they charge per thousand on whatever type of lumber you need.

Estimate your roofing needs by calculating the number of square feet in your roof (length times width on all slopes). Then give that figure to the counterman at the roofing material outlet, and he will tell you how much you need, depending on your roofing material.

### Estimating Roofing and Siding

Estimating the amount of siding depends on the material. If it's plywood, you know it comes in standard 4-by-8-foot sheets, so just count up what you need from your plans. If the siding is shiplap or boards and battens, count how many board feet would be required to cover an 8-foot section of the wall and then convert that to an estimate of your total needs.

### Selecting the Best Wood for Painting*

The degree to which wood accepts and holds paint is affected by its density and texture; its resin, oil, and moisture content; and defects such as knots. Generally, the more porous the wood, the better it takes paint. Pine, therefore, is one of the best choices for painting, and oak is one of the poorest choices. The resins and oils in wood can shorten the life of paint that is applied, but this problem can be effectively overcome by applying a primer before painting. The greener the wood, the more oils and resins it will contain. Freshly cut lumber should not be painted.

Similarly, a high moisture content will cause the paint to blister and peel. Boards to be painted should be stamped S-DRY, which indicates the moisture content is 19 percent or less.

Knotty lumber is a poor choice because the brown substances in knots will discolor the paint, and because knots do not hold paint well and may crack, rupturing the surface of the dried paint.

The best wood species to paint break down into five categories, in a sequence of descending quality:

**1.** Woods on which paints of the widest range in kind and quality give good service: cedar, bald cypress, and redwood. (Cedar and redwood, however, are rarely painted.)

**2.** Woods that require more care in selecting a suitable priming paint, such as a zincless house-paint primer: eastern white pine, sugar pine, and western white pine.

**3.** Woods that are more exacting than those in group 2 but less exacting than those in group 4 in terms of selecting the proper primer and paint: white fir, hemlock, ponderosa pine, and spruce.

**4.** Woods that require the most care in selecting the best primers and paints: Douglas fir, southern yellow pine, and western larch. Hardwoods included in this category include beech, birch, gum, and maple.

**5.** Woods that require wood filler in the pores before a smooth coating of paint can be applied are all hardwoods: ash, chestnut, elm, oak, and walnut.

*Courtesy of the U.S. Department of Agriculture

# CHARACTERISTICS OF WOOD

## Characteristics of Wood

These charts give detailed information on softwoods and hardwoods. The softwoods are used in rough carpentry while the hardwoods are used more often in fine woodworking. For the construction process described in this

### Softwood

| Type | Sources | Uses | Characteristics |
|---|---|---|---|
| Cedar (Western red, white) | Pacific Coast, Northwest, Lake States and Northeastern United States | Paneling, fenceposts, siding, decks. Red cedar used for chests and closet lining. | Fine grain; soft. Red: reddish-brown with white sapwood. White-light with light brown heartwood. Brittle, lightweight; easily worked; low shrinkage; high resistance to decay. Strong, aromatic. |
| Cedar (Red) | East of Colorado and north of Florida | Mothproof chests, lining for linen closets, sills. | Very light, soft, weak, brittle. Natural color. Generally knotty, beautiful when finished in natural color, easily worked. |
| Cypress | Southeastern coast of the United States | Interior wall paneling and exterior construction, i.e., posts, fences, cooperage, docks, bridges, greenhouses, water towers, tanks, boats, and river pilings. | Lightweight, soft, not strong, easy to work. Coarse texture. Durable against water decay. Light brown to nearly black. |
| Fir (Douglas) | Washington, Oregon, California | Construction flooring, doors, plywood, low-priced interior and exterior trim. | Light tan; moderately hard; close-grained. Most plentiful wood in the United States; used mostly for buildings and structural purposes; strong, moderately heavy. |
| Fir (White) | Idaho, California | Small home construction. | Soft; close, straight-grained; white with reddish tinge. Low strength; nonresinous; easily worked; low decay resistance. |
| Hemlock | Pacific Coast, western states | Construction lumber; pulpwood; containers; plywood core stock. | Light in weight; moderately hard; light reddish-brown with a slight purple cast. |
| Larch | East Washington, Idaho, Oregon, Montana | Framing, shelving, fencing, shop projects, furniture. | Moderately strong and hard. Glossy russet colored hardwood with straw colored sapwood. If this wood is preservative treated, it can be used for decking. |
| Pine (Lodgepole) | West Coast from the Yukon to Mexico | Framing, shelving, fencing boards, small furniture, shop projects. | Hard, stiff, straight-grained. Mills smoothly, works easily, glues well, resists splintering, and holds nails well. Light brown heartwood, tinged with red with white sapwood. |
| Pine (White, Ponderosa) | United States | Solid construction in inexpensive furniture, sash, frames, and knotty paneling. | Soft; pale yellow to white in color; fine-grained; darkens with age. Uniform texture and straight grain; lightweight, low strength, easily worked; has moderately small shrinkage, polishes well; warps or swells little. |
| Pine (Southern) | Southeastern United States | Floors, trusses, laminated beams, furniture frames, shelving. | Strongest of all softwoods with a pale yellow sapwood and reddish heartwood. |
| Pine (Sugar) | Western Oregon's Cascade mountains, Sierra Nevadas of California | Shingles, interior finish, foundry patterns, models for metal castings, sash and door construction and quality millwork. | Lightweight, uniform texture, soft. Heartwood is light brown with tiny resin canals that appear as brown flecks. The sapwood is creamy white. Straight-grained and warp resistant. |
| Redwood | West Coast | Sash, doors, frames, siding, interior and exterior finish, paneling, decks. | Heartwood is cherry to dark brown; sapwood is almost white. Close, straight grain. Moderately lightweight, moderately strong; great resistance to decay; low shrinkage, easy to work, stays in place well; holds paint well. |
| Spruce | Various parts of the United States and Canada | Indoor work only. Pulpwood; light construction and carpentry work. | Soft, lightweight, pale with straight unpronounced grain and even texture. |

book, you will be concerned mainly with softwoods. You'll find the chart on hardwoods helpful for your other woodworking projects.

**Hardwood**

| Type | Sources | Uses | Characteristics |
|---|---|---|---|
| Ash (White) | United States | Upholstered furniture frames, interior trim. | Hard; prominent, coarse grain; light brown. Strong, straight grain; stiff, shock resistant; moderate weight; retains shape, wears well; easily worked. |
| Beech | Eastern United States | Flooring, chairs, drawer interiors. | Hard; fine grain; color varies from pale brown to deep reddish-brown. Heavy, strong; has uniform texture; resists abrasion and shock; medium luster. |
| Birch (Yellow, Sweet) | Eastern and Northeastern United States and Lake States | Cabinet wood, flooring, plywood paneling, exposed parts and frames of furniture. | Hard, fine grain; light tan to reddish-brown. Yellow is most abundant and important commercially; white sapwood and reddish-brown heartwood. Heavy, stiff, strong; good shock resistance, uniform texture; takes natural finish well; satiny in appearance. |
| Cherry (Black, wild) | Eastern and Northern United States | Paneling, furniture. | Moderately hard; light to dark reddish-brown, fine grain; darkens with age. Strong, stiff, heavy; high resistance to shock and denting; not easily worked; high luster. |
| Gum (Red, Sap) | South | Plywood, interior trim, posts, stretchers, frames, supports — frequently used in combination with other woods. | Heartwood (red gum) is light to deep reddish-brown; sapwood (sapgum) nearly white; moderately hard, fine grain. Moderately heavy, strong, uniform texture; takes finish well; frequently finished in imitation of other woods. Fifty years ago, it was the most frequently used furniture wood in this country. |
| Hickory | Arkansas, Ohio, Tennessee, Kentucky | Tool handles, wagon stock, baskets, wagon spokes, pallets, ladders, athletic goods. | Very heavy, hard, stronger and tougher than other native woods; sapwood and heartwood of same weight. Difficult to work, subject to decay and insect attack. |
| Maple (Sugar, Black) | Great Lakes, Northeast, Appalachians | Interior trim, furniture, floors in homes, dance halls, bowling alleys. | One of America's hardest woods; heartwood is reddish-brown; sapwood is white; usually fine, straight-grained; sometimes curly, wavy, or bird's eye grain occurs. Strong, stiff; good shock resistance; great resistance to abrasive wear, one of the most substantial cabinet woods; curly maple prized for fiddle backs. |
| Oak (Red, White) | Eastern United States, mainly Mississippi Valley and South. | Flooring, interior trim, furniture, plywood for cabinet work, paneling. | Hard, pronounced open grain; rich golden color to light reddish-brown. Moderately heavy, stiff, strong, resilient, tough; comparatively easy to work with tools; takes many finishes. |
| Poplar (Yellow) | Eastern United States | Interior trim, siding, furniture, panels, plywood cores. | Sapwood is white; heartwood is yellowish-brown tinged with green; soft; straight, fine grain. Lightweight, moderately weak, does not split readily when nailed; easily worked, easy to glue; stays in place well, holds paint and enamel well; finishes smoothly. |
| Sycamore | Eastern half of United States | Interior trim, fancy paneling, furniture. | Light to reddish-brown; hard, close, interlocked grain. Moderately heavy, strong; rays are conspicuous when quarter-sawed; seasoning without warping is difficult. |
| Walnut (Black) | Central United States | Furniture, paneling, cabinet work, interior. | Light to dark chocolate brown; hard; moderately prominent, straight grain; sapwood is nearly white. Strong; resists shock and denting, easily worked; takes stain and finishes exceedingly well; heavy, stiff, is stable in use; one of most beautiful native woods; has luminous finish. |

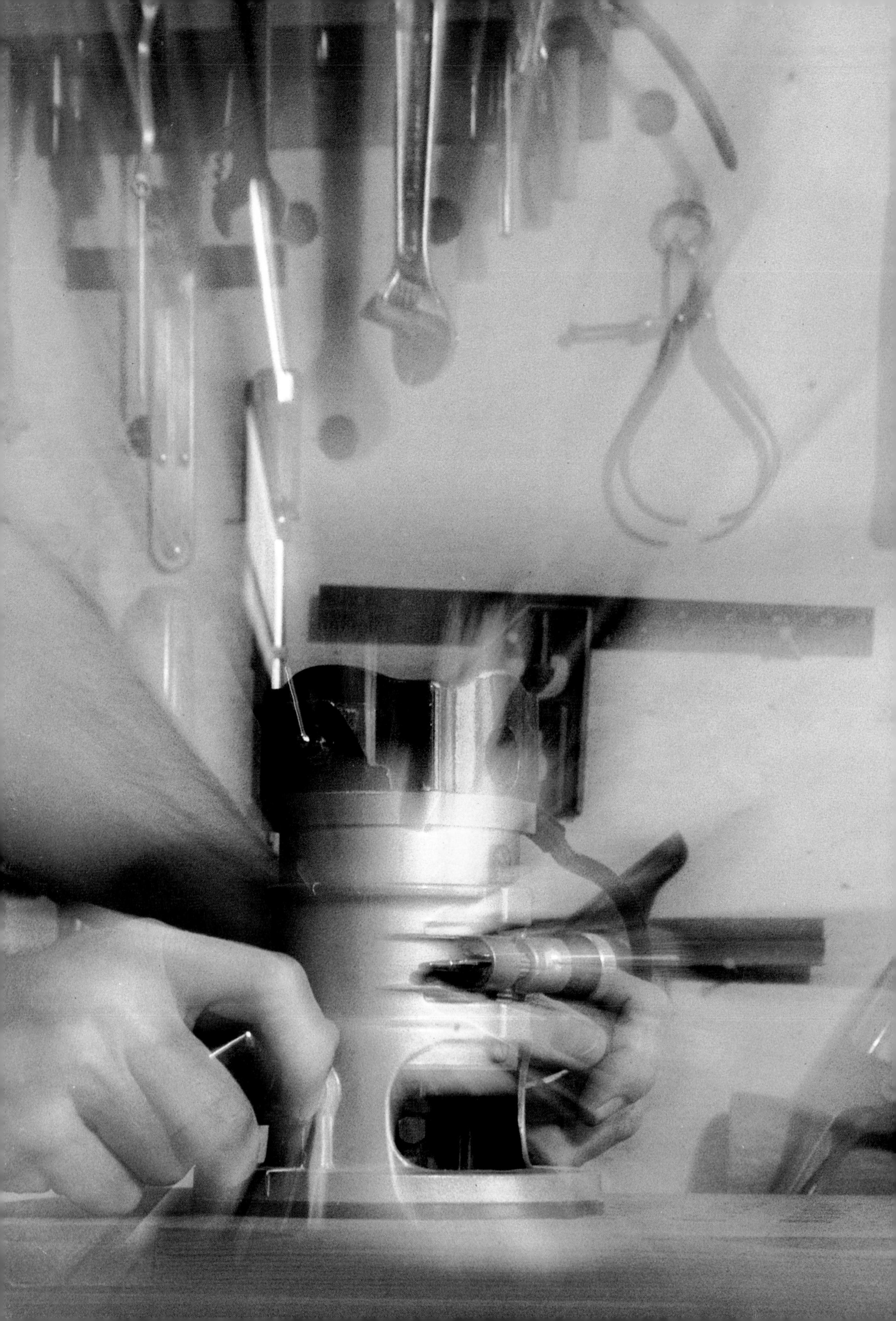

# TOOLS AND TECHNIQUES

From hammering a nail to making the perfect wood joint, every carpenter needs to know which tool to use and how to use it properly. This survey of the most widely used hand and power tools illustrates all the basic carpentry techniques and includes handy tips from the professionals.

## The Joy of Tools

The wonder of tools is that so much can be done with so little. A house can be built with just a hammer, nails, saw, square, tape measure, and hand level. The same is true of smaller projects such as bookshelves, drawers, and cabinets.

Fortunately there are a number of power and hand tools available today that make carpentry projects easier. Purists decry the extensive use of power tools, and to some degree they are right. Carpenters should know the fundamentals of using the variety of handsaws available. They should know the feel and smell of a sharp plane peeling away a ribbon of wood to make a door fit exactly, or the bite of a well-honed chisel trimming a piece of wood.

Power tools have their place in a workshop, but their use must be kept in perspective. They speed up the job of carpentering; they do not always improve the workmanship. And good workmanship must be your constant goal.

Work with the hand tools described in this chapter as much as you can. They may seem awkward and slow at first, but it's only through experience that you will grow skilled and confident with them.

This chapter is designed to make you aware of the most widely used tools in construction carpentry, and exactly what each of them can do for you. For example, you will learn which hammer to use to remove a bent nail, how to use it, and how to use an even handier little gadget called a nail puller.

This chapter also deals with the care and handling of tools. Tools in good condition are the mark of a good carpenter. If the tools are rusty, bent, and dull, you can assume the carpenter doesn't care much about workmanship either. Keep your own tools clean, oiled, and sharp.

◀

**The versatile router can be used to make fancy edges, cut dovetails, make moldings, and trim laminates. See pages 44–45.**

The importance of buying only the best-quality tools cannot be overemphasized. You'll do far better with a few fine tools than with many cheap ones. You are the best judge of which tools you will need. If you are just starting out you will need a hammer, saw, combination square, drill, and a couple of different screwdrivers. With that moderate investment you can cut boards that are square on the ends and put them together to make a simple bookcase or a workbench for your garage.

When you want more tools, don't buy on impulse; keep a running list of what you really need. And it's easier on the budget to buy tools one at a time.

**Construction work sometimes requires drilling into concrete. For details on this and all types of drills and techniques, see pages 40–43.**

## Types of Hammers

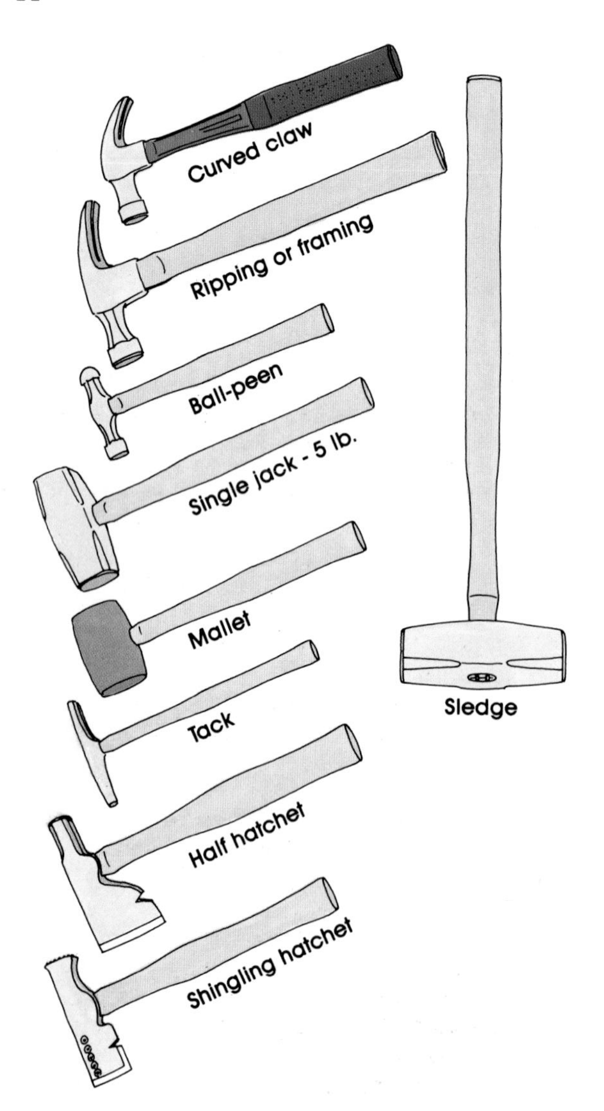


Any workshop, even if it is in a closet or a kitchen drawer, needs two or three different types of hammers. Hammers come in a wide selection of weights, which are stamped in ounces on the hammer head, and in different styles for different jobs. A survey of basic hammers is presented here.

### Curved Claw Hammer

This is the basic hammer for a shop. For general work it is usually 16 ounces in weight. It has a curved claw that permits a rocking motion for pulling nails with minimum damage to the wood.

### Ripping Hammer

This is the basic hammer for construction carpentry. Also called a framing hammer, it has claws that allow you to pry apart boards. Nails are easily pulled with a side-to-side motion (see page 19). Weights commonly run from 20 to 28 ounces.

### Singlejack Hammer

This is often called a five-pound hammer. It is handy to have around the shop and construction site, since it is manageable yet packs a big wallop. It works well for tasks such as hammering large beams into place.

### Sledgehammer

Weighing up to 20 pounds, sledgehammers are designed for heavy-duty service, such as knocking beams into or out of place, knocking walls apart, and driving wedges.

### Ball-Peen Hammer

The primary use for this hammer is metalworking, but the ball-peen is valuable in a shop because the head is made from specially hardened steel. For example, it can be used when you chisel into concrete, reducing the chance of chipping the hammer head.

### Mallet

This comes with a rubber, leather, or plastic head. Use it when you need weight but want to avoid hammer damage—for example, when hammering wood joints together. It is also commonly used to drive wood chisels.

### Tack Hammer

This has a magnetic head to hold tacks in place and is always a useful addition to a shop.

### Half-Hatchet

The half-hatchet is designed for rough carpentry work. The hammer end will drive nails and the cutting end will chip or split wood when necessary, to make stakes, and to hammer nails in a corner.

### Shingling Hatchet

You can use a shingling hatchet to put on a shingle or shake roof. A small plug fits in the holes to provide a measuring gauge for spacing the overlap of shingles, and the hatchet end allows you to split shingles when necessary. For details see page 100.

## Hammer Handles

A good handle not only adds power and speed to your striking blow but reduces any shock that might be transferred to your hand. Which type you choose depends mostly on personal preference.

**Hickory Handles.** These have been around the longest and are valued for their strength, shock-absorbing qualities, and the natural feel of the wood. They are fixed to the head by one or two wedges driven into the end grain. This splits the wood slightly, forcing it to press tightly against the opening in the hammer head.

**Tubular Steel Handles.** These handles are mechanically fixed to the head. They often bend if you strike the handle on a large nail or the edge of a board.

**Solid Steel Handles.** Forged in one piece with the head, these handles are especially strong. Like tubular steel handles, they usually have rubber or leather grips.

**Fiberglass Handles.** Fiberglass handles provide great strength and good shock-absorbing qualities. Some types are quickly and easily replaced if broken; others are held in place by epoxy glue and are more difficult to replace.

## Hammer Faces

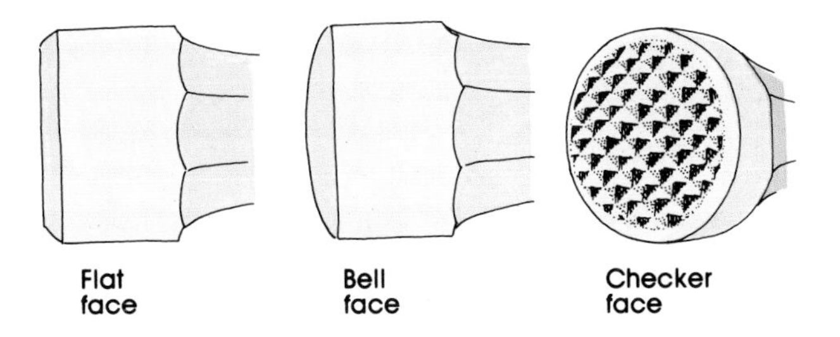


Hammers have different faces that are designed to do different jobs. For example, the standard flat face is designed for general household use, the bell face for finish work, and the checked face to help prevent slipping when the nail is struck. Different faces can be found in different styles. In other words, you can find a bell face in both the curved claw and straight claw hammer.

**Flat-Faced Hammer.** This hammer is designed to hit the nail as squarely as possible and should be the first choice for beginning carpenters. In skilled hands it will drive nails below the surface without leaving a mark on the wood.

**Bell-Faced Hammer.** The face of this hammer has a slightly protruding, or convex, surface. Its chief advantage is that it allows you to drive the nail exactly flush or slightly below the surface without marring the wood. It takes practice and skill to do this consistently. This hammer is most commonly used in finish or interior work.

**Checker-Faced Hammer.** Generally, this face is found on ripping (framing) hammers, where speed and accuracy are important. The broad flat face is cross-checked to minimize chances of glancing off the nail.

## Buying a Hammer

There are several elements to consider when you buy a hammer. The first rule is to buy the very best quality. Second, get the right hammer for the job. If you are driving tacks, don't choose a big framing hammer unless you have taken a strong dislike to your fingers.

To determine the quality of a hammer, first make sure that the head is made of forged steel. In quality hammers, this will be stamped on the head. Don't buy cast iron, which is much weaker and may break. Also, buy a hammer with a solid handle rather than one made of light tubular steel, which bends easily. The cheaper hammers may be all right for putting a nail in plasterboard to hang a picture, but not much beyond that. If in doubt ask the hardware store salesperson.

Choose a hammer that is not only the right style for the job but the right weight, as well. They range from 7 to 38 ounces, and the weight is normally stamped on the cheek of the face. For little jobs around the kitchen use a curved claw hammer of 7 to 12 ounces. The 16-ounce curved claw hammer with a bell face is a good choice for use in the shop. It can readily pull small nails, its moderate weight makes it ideal for a wide variety of tasks, and the bell face allows you to do finish work without marring the wood.

The next important hammer to buy for your shop is the framing hammer, used for heavier tasks. The most commonly chosen weights are 20 to 24 ounces, depending on the strength of your wrist. When you are selecting a framing hammer, swing it a few times and see which weight feels the most comfortable.

The striking surfaces and the top of the claws should be smooth to minimize marring, but not necessarily polished, since polishing adds more to the cost than to the quality of a hammer. Look for a smooth and sharp V where the claws meet so you can get a firm hold on the shank of a nail. You don't want to hook right under the nail head, since it may pull off.

When you are checking a hickory handle, look at the end grain. On the strongest types the grain will run parallel to the head.

## Replacing a Handle

Wood handles are held tightly in the hammer head by one or two small wedges driven into the end grain of the handle. This forces the wood apart slightly and thus binds the handle to the head. But handles can loosen when a wedge falls out or the wood shrinks.

Wedges for wood hammer handles are available in most hardware stores. Before replacing a handle, try hammering one or two into the top end of the handle. Always place them so they are perpendicular to the hammer head.

If additional wedges will no longer keep the head on tight, or if the handle becomes splintered, it's time for a new handle. Cut the old handle just below the head and drive the piece in the head out of the hole. Sometimes this can be quite difficult. If you can't drive it out, try splitting it with a narrow chisel and working the pieces out. If all else fails, burn it out with a propane torch.

Shape the new handle with a wood rasp for a tight fit in the head and then saw a slot two-thirds the head's depth

**Parts of a Hammer**

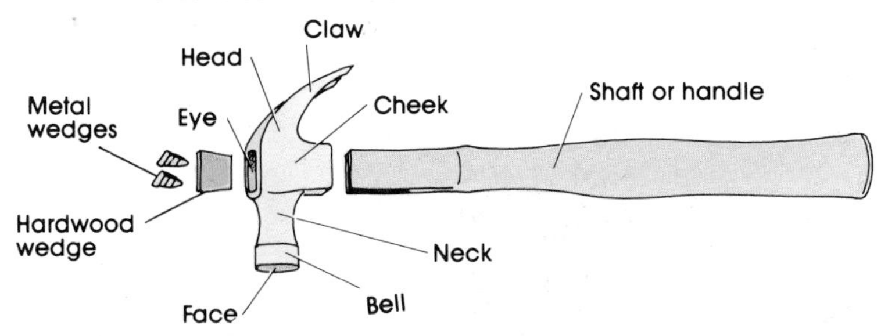


**Replacing a Handle**

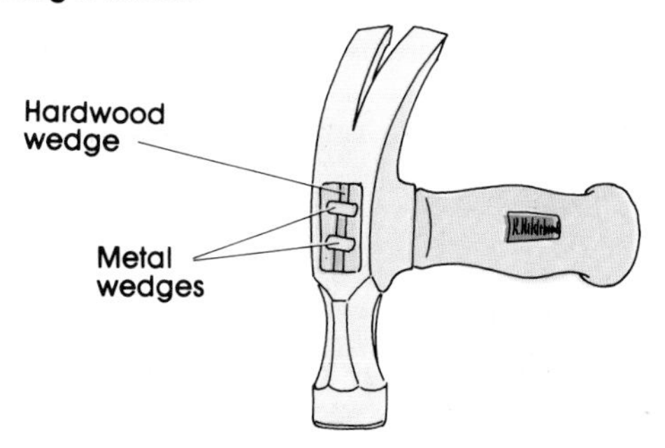

vertically down the center of the handle.

Drive the new handle into the head by banging the base of the handle on a hard surface until you have a smooth, tight fit. Drive a hardwood wedge into the slot as far as you can. Cut the protruding end piece flush with the head and then drive two metal wedges in perpendicular to the hardwood wedge. File the end smooth and give it a coat of varnish. Give the handle a light sanding and rub with boiled linseed oil.

### Hammer Maintenance

Treat a hammer like the good friend it is. Inspect it regularly for damage. If the face is chipped, you should replace it. Not only can it result in poor nailing, but it may chip again, which is a danger to you and your fellow workers.

Don't keep the hammer in an area of high humidity. Humidity causes the wood to swell around the base of the head, crushing the fibers. This weakens the handle and will loosen it if it dries out.

Don't keep a hammer in an overly dry or hot area, such as above a radiator or a stove. This will cause the wood to dry and shrink. If this does happen, put it in a bucket of water for an hour to swell the wood back and then keep it at a moderate temperature.

Don't use nailing hammers on steel; use a specially hardened ball-peen hammer. Wipe the head with an oily rag when you are finished to prevent rust.

## Techniques for Hammering and Pulling Nails

Even the best of carpenters are not immune from *fingersmashitis.* Since it is certain that you will hit your fingers many times, you have only two consolations: you are in good company, and the more practiced you become, the less of a risk you will be to yourself.

### Starting Nails

Start by holding the nail between your thumb and index finger near its head. Do not let your fingers rest on the wood. If you do, and you miss, you will drive your finger directly against the wood. Tap the nail head lightly but firmly once or twice at most to set it in the wood. Then hold the piece of wood at a safe distance away from the nail and drive the nail with smooth, steady strokes.

For nails that are too small to hold between your thumb and index finger, place your hand flat on the board and slip the nail between your index and middle finger. Hammer very carefully.

### Striking Nails

Always make sure that you strike the nail with a blow that is exactly vertical. If the nail bends, either you have run into a knot or you are hitting it at an angle. To test your accuracy, lay a piece of scrap board on a workbench and make a series of pencil dots about three inches apart. Hit each one in succession. Now study the indentations to see how accurately you are swinging. The indentations should be centered over each pencil dot. Just as important, each indentation should be an even depth all around. If they aren't, you can see that you are tilting the hammer face slightly one way or the other, delivering a glancing blow that bends the nail.

### Proper Swing Sequences

Swinging the hammer correctly involves a smooth coordination of wrist, elbow, and shoulder. Once the nail is started, pull your fingers back and lightly touch the hammer to the nail head. This allows you to "sight" your target. Draw the hammer back, flexing your wrist up a little; then hit the nail.

To drive small nails accurately, use mainly wrist action.

You will need a heavier swing to drive medium-sized nails (6d to 10d). (See page 20 for comments on nail sizes.) Elbow movement becomes important, giving more authority to your wrist action.

With larger nails it takes a full combination of shoulder, elbow, and wrist action—as well as a big enough hammer. Again, the swing must be one fluid motion, with the hammer coming down squarely on the nail.

One sign of poor hammering technique is the series of little quarter moons in the wood around the nail head. You may have driven the nail correctly, but don't ruin the job by repeatedly smashing the wood around the nail. As you become more experienced, you can accurately sense when the last blow will drive the nail flush with or below the wood without leaving a single mark.

**Techniques for Hammering**

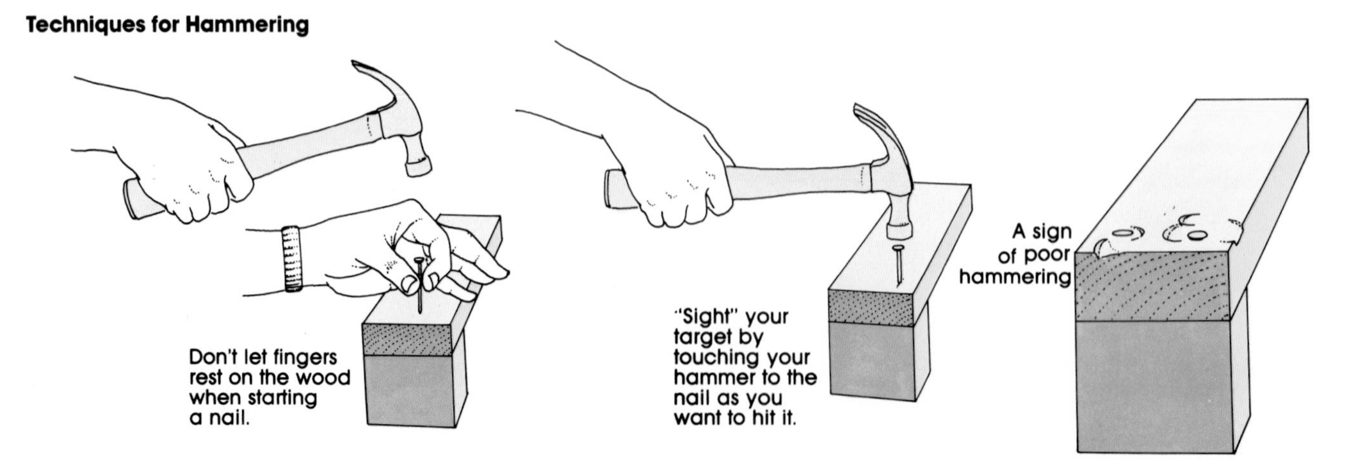

## Toenailing

Toenailing (shown in the illustration) is a fundamental technique of carpentry. It is primarily used when it is impossible to drive nails through one board into the end of another board. A common example of toenailing in house construction is where studs must be nailed to a sill board that is already fastened to a concrete foundation.

The key to proper toenailing is to start the nails high enough on the wood so as not to split it, but low enough to get a good bite in the lower piece of wood. The nails should be at about a 60° angle. Use at least two nails on each side for larger pieces of wood.

When possible, brace one side of the upright piece of lumber with your foot to prevent it from shifting while you drive the nail. Try not to nail your shoe to the floor in the process.

## Placing Nails

Some building codes contain nailing schedules, but in general, when you nail down a board, use just enough nails to do the job. You will gain a sense of just how many nails to use as you become more experienced. Keep in mind that boards tend to work loose at the ends first, so you will need at least two nails on each end to keep a 2 by 4 or 2 by 6 flat.

As you nail along a board, do not place the nails in a straight line. This is akin to driving in a series of tiny wedges and can split the board along the grain. Always stagger the nails.

When you nail close to the end of a board, it often splits. Here's a trick to counter that: tap the nail lightly on a hard surface to flatten the point a little. When you drive it through the wood, it will tend to punch its way through the wood fibers rather than forcing them apart into a split.

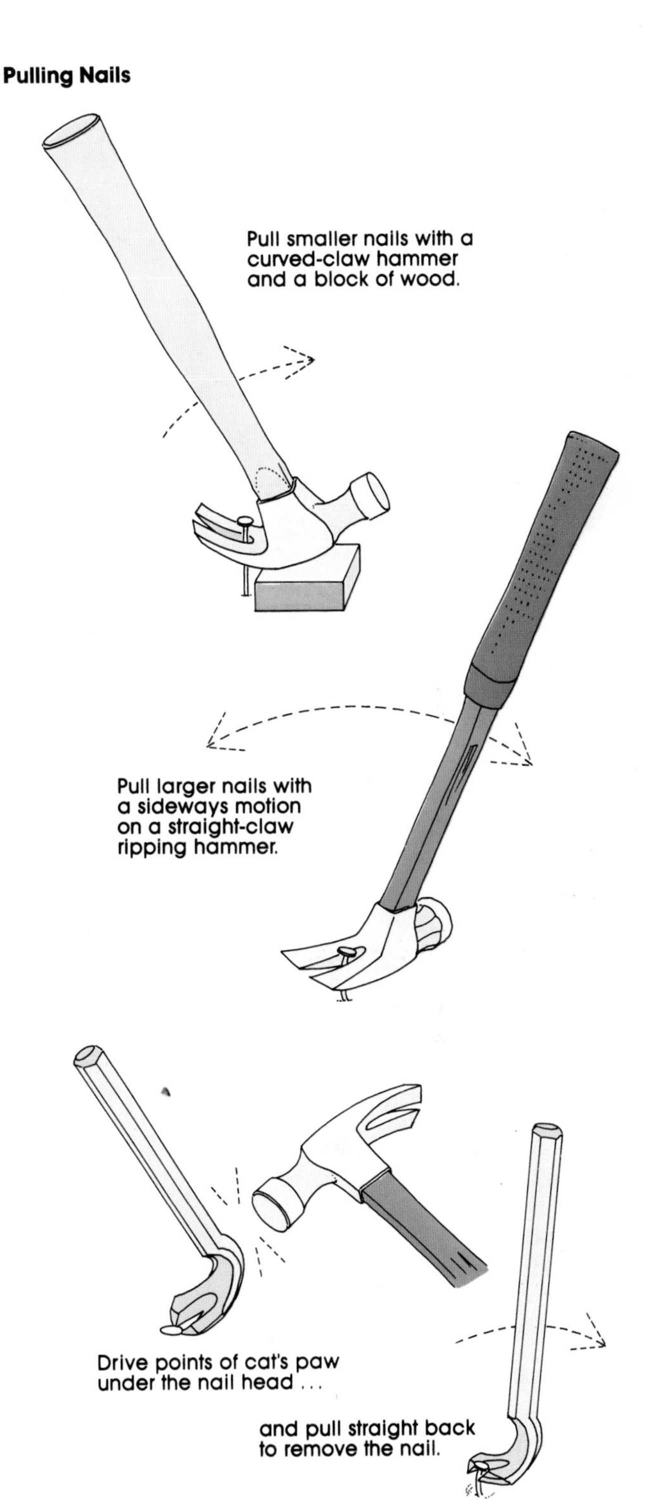


## Pulling Nails

Small nails can be pulled with a curved claw hammer. For additional leverage, place the hammer on a block as shown. This also prevents the hammer from marring the wood. Hook the V of the claws on the shank of the nail, rather than just under the nail's head, which may pull off.

For larger nails (8d and up), there is a trick that usually makes nail pulling easy. Use the straight claw ripping hammer for this. Hook the nail by the shank and then push the hammer over to one side. Hook it again and push the hammer in the opposite direction. The biggest nails can be easily worked out in this way. When it is necessary to prevent marring, place a piece of plywood or thin wood between the hammer and wood.

Nails that are driven completely flush with the wood present more of a problem. The easiest way to deal with them is by using a nail puller, sometimes called a cat's paw. Hold the points of the small claws just back of the nail head and then use your hammer to drive the points down and under the nail head to hook it.

If you don't have a nail puller, you can accomplish the same thing by driving the straight claws of a ripping hammer around the nail. Curved claw hammers cannot easily be used for this task. However, do not drive the claws under a nail by tapping the face of the hammer with another hammer. You will chip the face, which can be dangerous.

# NAILS

**Types of Nails**

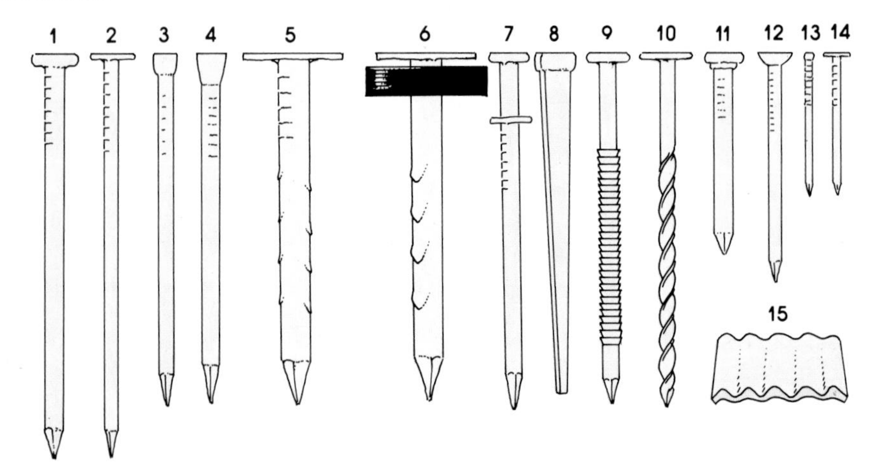


1. Common
2. Box
3. Finish
4. Casing
5. Galvanized roofing
6. Aluminum roofing
7. Scaffold
8. Cut floor
9. Ringed
10. Spiral
11. Concrete
12. Drywall
13. Wire brad
14. Flat head wire brad
15. Corrugated fastener

There are five basic types of nails: common, box, coated, finish, and casing. Common and box nails (both coated and uncoated) are the standard construction nails, while finish and casing nails are used for interior finish work. In addition, there are a number of specialty nails that are used for specific jobs.

Nail lengths are designated by the term *penny*, which is abbreviated "d." Thus a six-penny (6d) nail is always the same length (2 inches) no matter what type it is.

Nails are sold either by the pound or by the case. A case contains 50 pounds of nails. The five basic types of nails and a variety of specialty nails are described below.

**Common nails.** These nails are the first choice for construction work such as framing because of their strength, holding power, and great variety of sizes.

**Box nails.** These are exactly the same as common nails except that the diameter of the shank is smaller. This makes them somewhat more likely to bend when being driven, but the smaller shank is less likely to split wood. They are excellent for putting on siding.

**Coated nails.** These nails, in both common and box varieties, are coated with either a thin cement coating or vinyl. The coating melts slightly as the nail is driven in, then hardens to provide greater holding power. They are more expensive than uncoated nails but go in faster and hold better. They are a good choice for construction work.

**Finish nails.** The slender finish nail has a small, barrel-shaped head that can be driven beneath wood surfaces with a nail set (see page 60) and then covered with wood putty to hide it. They are good for window and door frames or paneling.

**Casing nails.** The head on this nail is tapered, which provides slightly more holding power than finish nails. Casing nails have a slightly larger shank than finish nails. They are also used wherever the nail needs to be hidden.

**Galvanized nails.** These are coated with zinc to prevent them from rusting. They come in either box or common sizes and are used where they will be exposed to the weather, such as on decks or roofing.

**Roofing nails.** These have large, flat heads designed to prevent the nail from working through soft roofing materials. They are normally galvanized.

**Aluminum roofing nails.** These are similar in appearance to the standard roofing nail but are used only on aluminum roofing or siding. Using a galvanized or steel nail in aluminum sets up a chemical reaction that causes the nails to corrode. Aluminum nails may have a rubber or plastic washer under the head to prevent water from leaking through the nail hole.

**Duplex nails.** The double heads on these nails make them easy to remove. They are used for putting up temporary material, such as bracing, scaffolds, or foundation forms.

**Cut flooring nails.** These have flat sides and blunt tips designed to minimize the chances of splitting hardwood flooring.

**Annular ring nails.** The rings on these nails provide tremendous holding power. They are commonly used for nailing down plywood subflooring to prevent it from working loose.

**Spiral nails.** The spiraled shank on these nails makes them turn like a screw when driven into wood, increasing their holding power. The fairly small head can be set below the wood surface. These are a good choice for nailing down a squeaky floorboard.

**Concrete nails.** These are made of case-hardened steel for maximum resistance to bending. Use them to nail up furring (wood strips that provide a nailing surface for paneling) on concrete or masonry walls.

**Drywall nails.** These nails have a broad, cupped head to resist pulling through soft drywall, or sheetrock. They normally come in either 4d or 6d sizes, depending on the sheetrock thickness. Use 6d for ½-inch sheetrock.

**Wire nails and brads.** Flat-headed wire nails and wire brads are small nails used for putting up thin paneling, molding, or parquet floors. They are sold by the quarter-pound or half-pound box in several different sizes and diameters. The box is labeled with a number and length in inches, such as 18–¾". The higher the number, the smaller the shanks.

**Corrugated fasteners.** These are used for holding together mitered or butt joints, such as on picture frames or wooden screen doors.

# BASIC SAW CUTS

Sawing is probably the most basic skill required of any carpenter. The degree of precision you achieve with a saw can make the difference between shoddy and excellent work. If you know how to make the basic cuts with each of the many types of saws, you are well on your way to expertise.

The basic cuts are actually quite simple. Except for the occasional circles and curves, every saw cut is simply a straight line. Depending on the project requirements, you can cut with the grain (rip cut) or across it (crosscut) or at an angle (miter or bevel cut); you can also cut all the way through the wood or only partially (as in a dado). And several of these basic cuts can be combined.

Examples of the basic cuts are illustrated on this page. The next section outlines the basic wood joints and how to cut them. Subsequent sections describe how to select and use a variety of handsaws and power saws.

**Basic Saw Cuts**

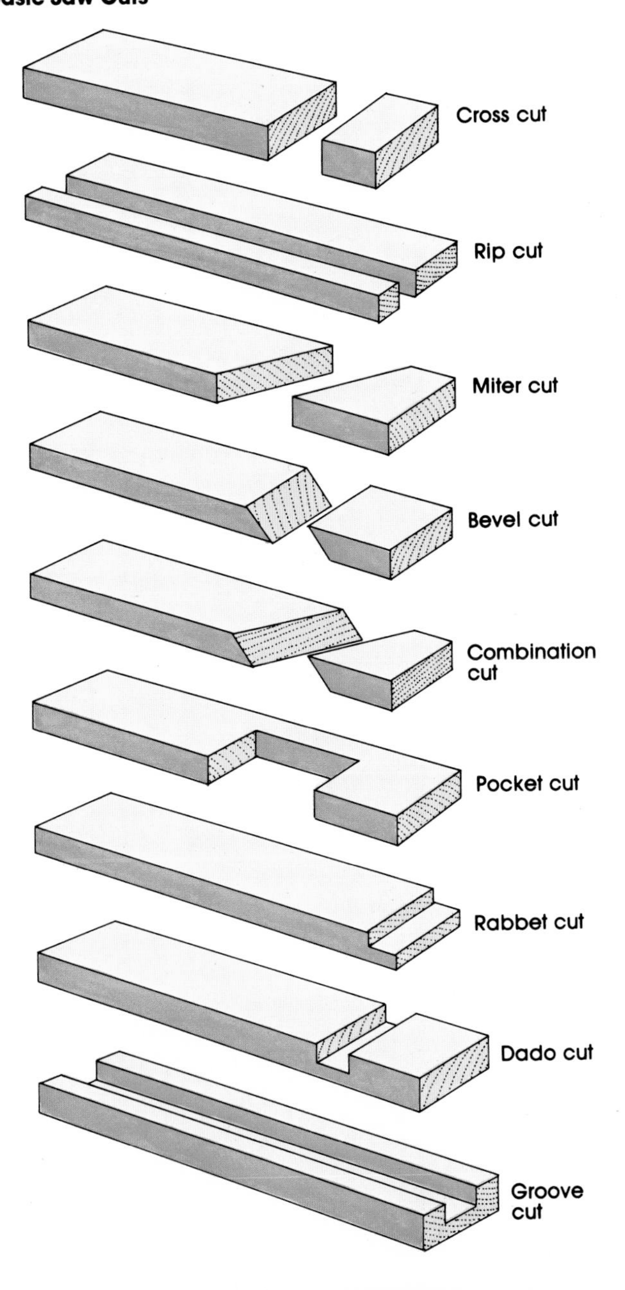


## Saw Cuts for Wood Joints

Many ingenious methods of joining two pieces of wood together have been devised over the centuries. Making a good wood joint requires skill and patience, but the results are worth it; the joint is both strong and beautiful. Putting wood together with some of the basic joints or angle cuts covered here can take you from the realm of a basic carpenter to that of a skilled cabinetmaker.

This section describes 11 joints that are basic to woodworking as well as various related cuts. The traditional way to make the cuts for these joints is with a handsaw (preferably a backsaw because it is quite rigid for precise work), a chisel, a drill, and a hammer. These techniques are illustrated here. However, all the cuts for joints can be made with power tools as well. While power saws will make the cuts more efficiently than a handsaw, many craftsmen believe they will not make as fine or accurate a cut. However, two types of cuts can be made with great accuracy on both table and radial arm saws: the dado and the groove. This is because both of these saws can use a special dado blade to make the cuts. The other power tool widely used for making dadoes, grooves, and rabbets is the router.

Instructions for using power saws to make the cuts described here are included with the sections on those saws later in this chapter. Choosing a power saw or a handsaw is really a matter of personal preference, but no matter what tools you choose, the cuts themselves and the joining techniques are the same.

## Overlap Joint

This is the simplest of all joints and involves little craftsmanship, although it is quite functional. It is made by cutting boards to length and laying one board on top of another, usually at right angles to each other, and then fastening them together. You can refine and strengthen this joint by using screws and glue instead of nails.

## Butt Joint

This is another simple and commonly used joint, but one with relatively little strength. When joining a butt end to a board face, cut the butt end perfectly square so it will fit smooth and tight against the face of the other board. Join

**Overlap Joint**

**Butt Joints**

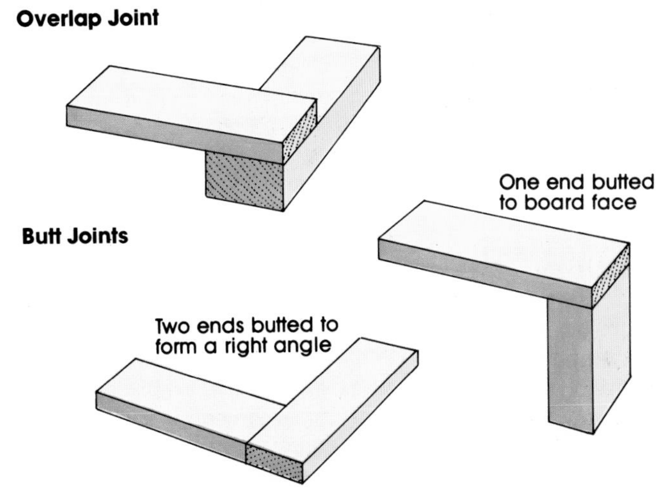

by nailing through the face of the board into the butt end, or by toenailing (see page 19).

If you want to butt two boards together at the ends to form a right angle, such as on a screen door, first put them together on the workbench. Next, drill two holes through the face of one board and into the butt of the other. Put glue in the joint and then screw it together. The screwheads can be hidden by countersinking them and covering them with wood putty. You may wish to use dowels instead of screws (see below).

## Doweled Joints

If you plan to make numerous doweled joints, consider buying a doweling jig. It usually comes with complete instructions and will allow you to do very precise work.

Use at least two dowels on any joint. In order to join two pieces of wood together smoothly, the dowel holes must be drilled in precise locations for a perfect match. The easiest method is called open doweling; a more difficult method is known as blind doweling.

*Open doweling* involves clamping or holding together the two pieces, then drilling through the outer edge of one board into the butt of the other. The hole in the butt end should be half again as deep as the connecting board is thick. The dowel should fit snugly, but not too tightly. Drill a hole in a piece of scrap and tap the dowel part way in to check the fit.

Bevel the end of the dowel slightly with a file so it will start into the hole smoothly, without splintering. The dowel must also be scored one or more times to allow air and excess glue to escape.

With the two pieces still clamped together after drilling the holes, squirt glue into the hole and then tap the dowel all the way in. Trim the excess dowel off flush with the outer edge and sand smooth.

### Dowel-Scoring Jig

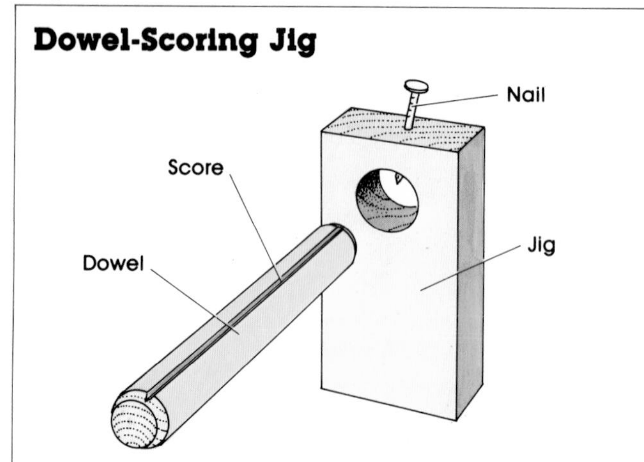


To score lengths of doweling quickly, drill a hole slightly larger than the dowel about an inch from the end of a scrap 2 by 4. Drive an 8d nail from the butt end of the scrap so the nail point just protrudes into the hole. As you hammer the dowel through the hole, the nail point will score the length of the dowel. Score each dowel at least twice to provide an escape channel for air and excess glue.

**Blind Dowel Joint**

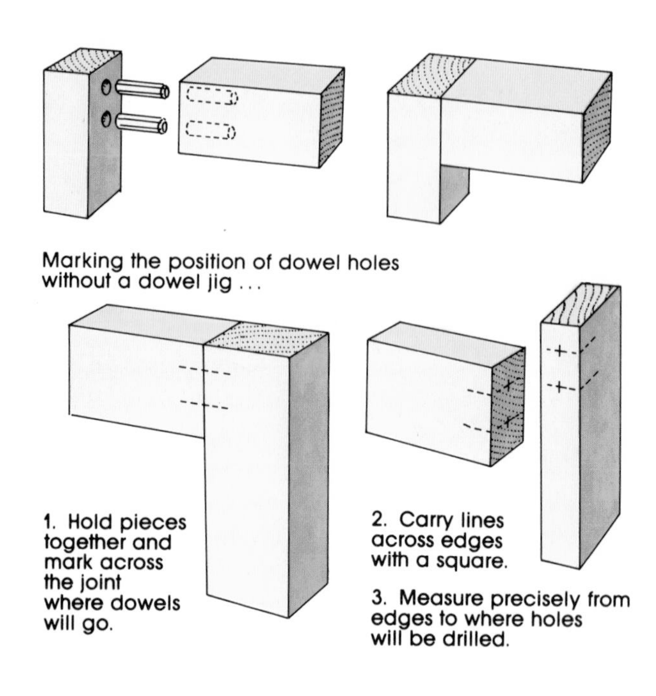


*Blind doweling* means joining two pieces of wood so that the dowels are invisible. To do this without a jig, clamp together the two pieces to be joined. Check that all edges are flush, then mark the dowel locations across the joint, on both pieces. Separate the two pieces and, with a square, carry these lines across the board edges to be joined. Measure precisely where the holes are to be drilled on each edge.

With a piece of tape on the drill bit to mark the proper depth, drill the dowel holes in each piece 1/16 inch deeper than the dowel length to provide room for glue at the rear of the holes.

Cut the dowels to length, drive them into the holes on one side, and then tap the other piece onto the exposed dowels. Do this first in a dry run to check for a flush fit before gluing. If your alignment is bad, fill the holes with dowels, cut the excess flush with the edges, and start over.

## Miter Joint

This joint, in which two pieces of wood, generally cut at 45 degree angles, fit together to form a right angle, is commonly used for picture frames, molding, and fine cabinet work. A 45 degree angle can be laid out on a piece of wood with your try square or combination square, both of which have such angles built into them (see page 59). When you cut freehand, mark the board both across the face and down the side so you can follow accurately with a backsaw.

You may wish to use a miter box rather than cutting freehand. The miter box can be either an inexpensive wooden one or a more elaborate (and expensive) metal one that adjusts to cut any angle between 30 and 90 degrees. In either case, place the wood to be cut in the box, hold it firmly with one hand, then put the backsaw in the 45 degree angle slot and cut.

**Miter Joint**

Hold pieces together with glue and finish nails, corrugated fasteners, screws, or dowels.

Two pieces of mitered wood can be joined in several ways. The simplest (and weakest) way is to glue both faces, pull them tightly together in a picture frame clamp (see page 46), and drive two corrugated fasteners across the joint on each side. A method that is nearly as simple but a little stronger is to glue the faces and nail them together. More strength is gained by drilling and then using glue and screws. For an even more professional appearance, put them together with dowels as described on page 22.

## Full-Lap Joint

This is a fairly simple but strong and attractive joint. The chief consideration here is that the board you cut must not be notched more than one-third its thickness; otherwise it will be too weak. Use a thicker board if necessary. To make the joint, place the lapping board in place on the cross-board and mark its position. Use a square to draw your lines. Measure the thickness of your lapping board and draw those measurements on the cross-boards, again using the square. Cut the sides of the notch and then use a chisel to break out the piece. Remember to hold the chisel with the flat side along the bottom line and the chisel's bevel facing the wood to be removed. Put the lapping board in the notch and fasten, usually with glue and screws.

## Half-Lap Joint

This is essentially the same as the full-lap joint, but a shallower notch is required because you also notch the lapping board. Start by measuring the thickness of the lapping board and draw a line through the center, using your square. Transfer this measurement—half the thickness of the lapping board—to the cross-board. Cut the cross-board notch. Then cut the lapping board notch. Fasten the joint together, usually with glue and screws.

Half-laps can be used very effectively to join boards in the middle, at right angles on the ends, or as an extension of each other. The latter use is appropriate when two large beams must meet over a supporting post.

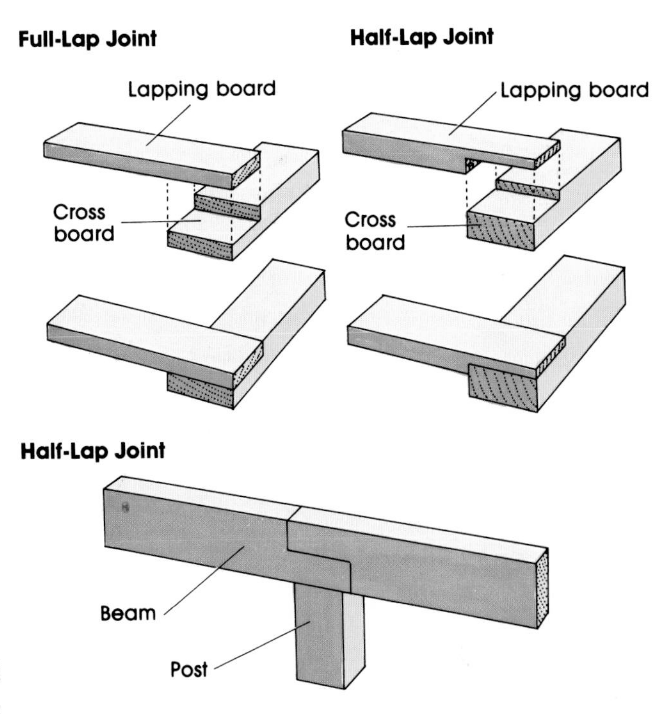


**Cutting a Lap Joint with Hand Tools**

## Plain Dado Joint

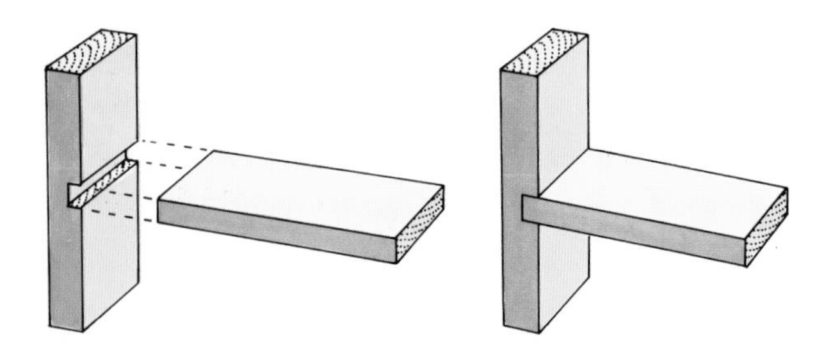

A *dado* cut goes across the grain of a board, whereas a *groove* goes with the grain. Dado joints are used to set the butt end of one board into the middle of another board. Dado cuts provide support, such as for shelving. They are best done with dado blades or a router, but a saw and chisel will suffice if you follow the instructions given here.

Using a square, draw two parallel lines across the face of the board to be dadoed, representing the exact thickness of the board to be inserted in the notch. Mark the depth of the cut required on the edges of the board and cut it. The depth of the cut should be no more than one-third the thickness of the board.

Then chisel out the dado, starting from one edge, with the chisel's bevel up, and work up toward the center of the board. Repeat this process from the other edge. Then turn the beveled edge down and remove the remaining waste.

## Groove

As stated above, a groove is made just like a dado, except that it runs with the grain rather than across it. Grooves are commonly used for drawer sides. If the groove is longer than your backsaw, clamp straight edges, such as 2 by 2s, along the line to be cut. This will help keep your saw directly on the line. Cut one line to the proper depth, then move the guide to the other line and cut it. Use a chisel to remove the waste wood.

**Cutting a Dado or Groove with Hand Tools**

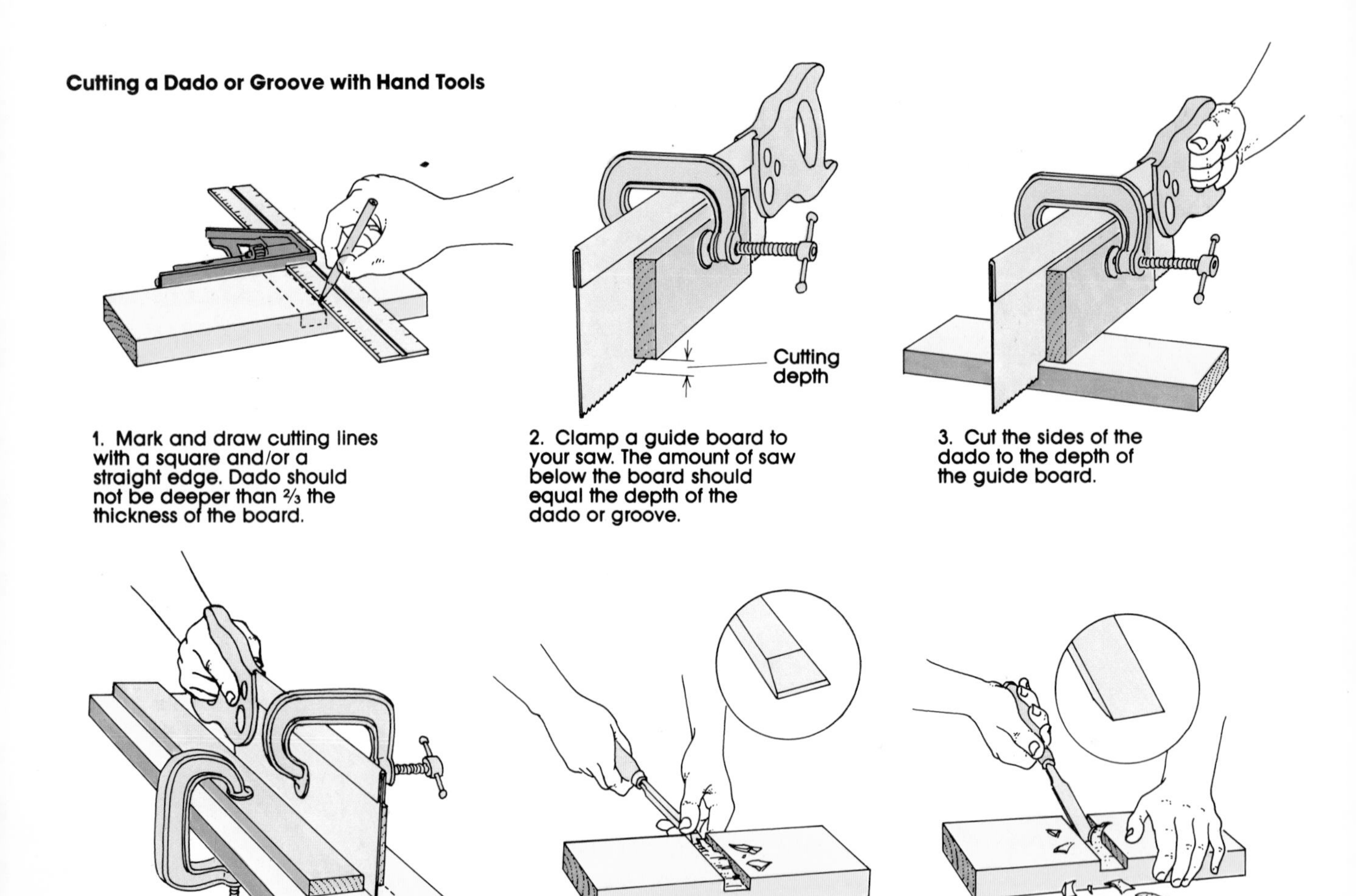


1. Mark and draw cutting lines with a square and/or a straight edge. Dado should not be deeper than ⅔ the thickness of the board.

2. Clamp a guide board to your saw. The amount of saw below the board should equal the depth of the dado or groove.

3. Cut the sides of the dado to the depth of the guide board.

4. The sides of a groove, even one longer than the saw, can be accurately cut by clamping another guide board to the work.

5. Remove most of the waste by working from each side at an upward angle toward the center. Hold the chisel with the bevel side up.

6. Clean out remaining waste and smooth bottom of dado or groove with the bevel side down.

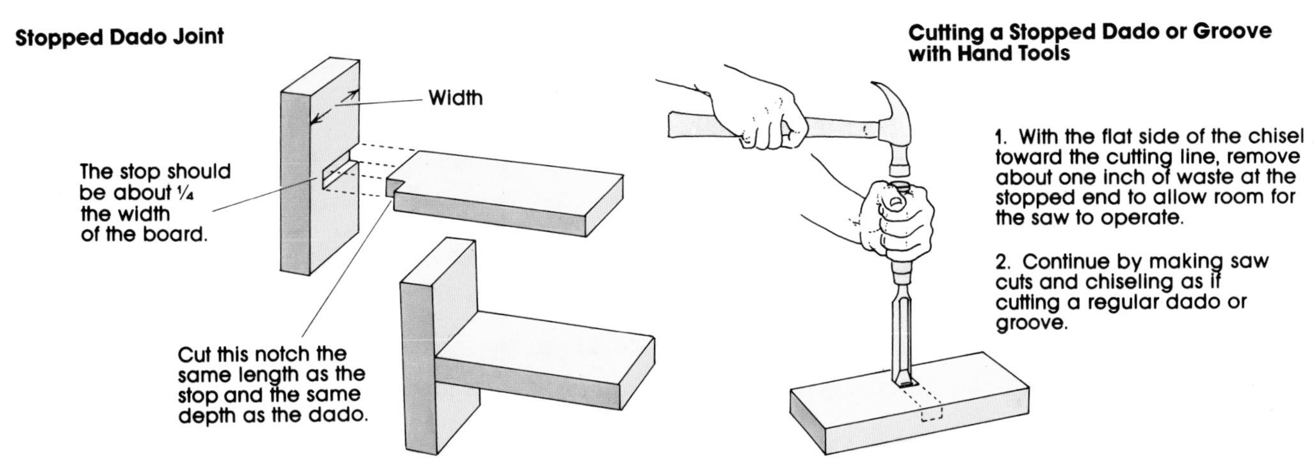


## Stopped Dado Joint

The stopped dado joint is frequently used for shelves so that you don't see the joint from the front. It is most easily done with a router or with a dado set on a table or radial arm saw, but you should also know how to make this joint with a handsaw.

First draw parallel lines on the board to be cut the exact thickness of the cross-member. The dado stops just a short distance from one edge of the board, about one-fourth of the board width. To make room for a handsaw to begin the cut, chisel out a small section near the stop. Saw this piece to the proper depth and chisel out the waste wood. Remember to keep the beveled side of the chisel facing the waste wood.

Measure the length of the stop on the supporting board and then cut a notch that size in the end of the cross-member. Apply glue to the end of the cross-member and slip it into place.

## Rabbet Joint

This joint is quite easy to make and provides a strong, finished appearance. It is an excellent choice for making drawers. Basically, it is a half-lap in one board that fits over the butt end of another. For adequate strength, the rabbet cut should be no deeper than three-quarters the thickness of the board. Use a square to mark the lines to be cut. First cut the depth of the rabbet; then, with the board held in a vise, make the vertical cut. Put the boards together with glue and finishing nails or screws.

## Mortise and Tenon Joint

This joint, commonly used in making furniture, is nearly invisible when done properly. Make the tenon first; the mortise will then be made to fit that tenon.

The tenon should be one-third the thickness of the board. Its depth is determined by the thickness of the board that is mortised. The mortise should be no more than two-thirds the thickness of the board. So measure the board that will be mortised and mark your tenon depth accordingly. Place the board in a vise, butt end up, and divide the butt into three equal sections. Carry these lines down both edges of the board the depth of the tenon. Using a backsaw, first make the two vertical cuts. Then place the board on your workbench and cut the board faces.

At this point you have a basic tenon, but its width still matches the board's width. You can add a little finesse to your work by trimming the two outer edges of the tenon so the shoulders all have the same depth. Measure the depth of the existing shoulders and then transfer these measurements to the tenon edges. Use a square when drawing the lines. Make the two vertical cuts first and then cut the shoulders.

For the mortise, place the butt of the tenon on the edge to be mortised and mark its outline with a sharp pencil. Alternatively, divide the edge to be mortised into three equal parts, with the center portion to be removed. When you need to mortise a door frame for a lock, measure the width of the bolt and mark the corresponding portion to be removed on the frame.

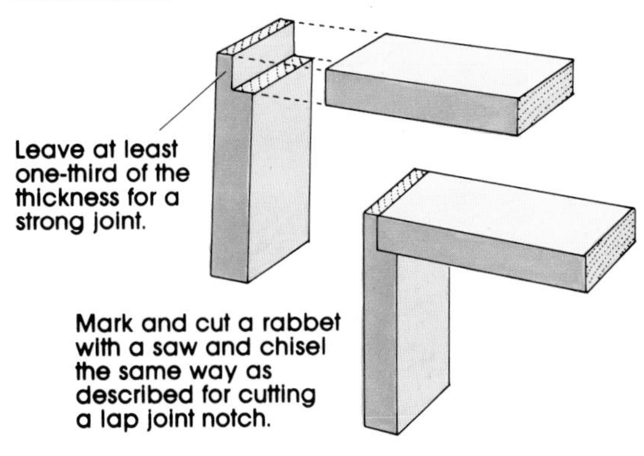


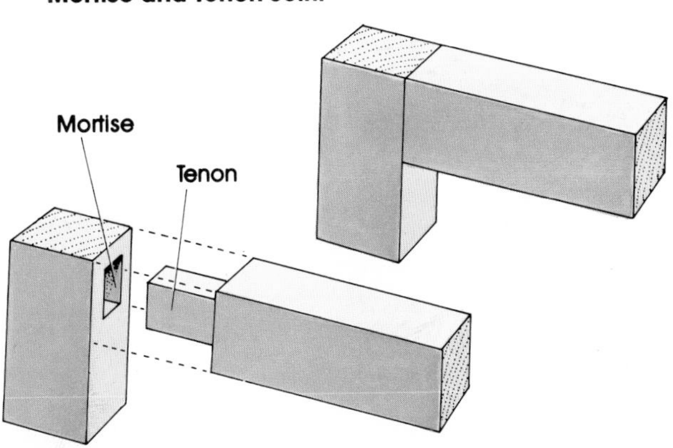

The smoothest means of removing the waste wood in a mortise is with a chisel. Work with the beveled edge toward the waste wood. Keep the edges vertical as you work down to the proper depth. Carefully measure the tenon to find the proper depth, then make the mortise 1/16 inch deeper to make room for glue that will be compressed at the back.

The mortise can also be cut quickly by drilling a series of holes. Put tape around the drill bit to mark the proper depth, and stop drilling when the tape reaches the board. The bit should be 1/16 inch smaller in diameter than the mortise opening. After the holes are drilled, use the chisel to square up the mortise edges.

Before applying glue, tap the tenon into the mortise with a mallet that will not mar the wood. When you have done any necessary trimming for a snug fit, apply glue to the butt end and sides of the tenon and tap it into the mortise.

## Open Mortise and Tenon Joint

This is a simpler version of the standard mortise and tenon, but equally strong. The tenon is made as described above. The mortise is easier to make because most of your work is done with a saw. Place the piece to be mortised in a vise, butt end up, and divide it into three equal sections as if you were making a tenon. Cut with a backsaw down the two vertical lines to the depth of the tenon. Now chisel out the center portion. Keep the beveled edge of the chisel toward the waste wood. Turn the board frequently so you work from both sides toward the middle. This will help you keep the base of the mortise flat. Tap the pieces together in a dry run to check for a smooth fit before applying glue.

**Open Mortise and Tenon Joint**

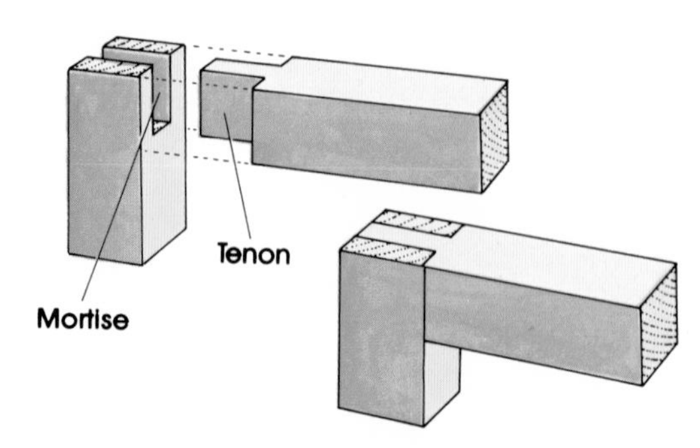


**Cutting a Mortise and Tenon with Hand Tools**

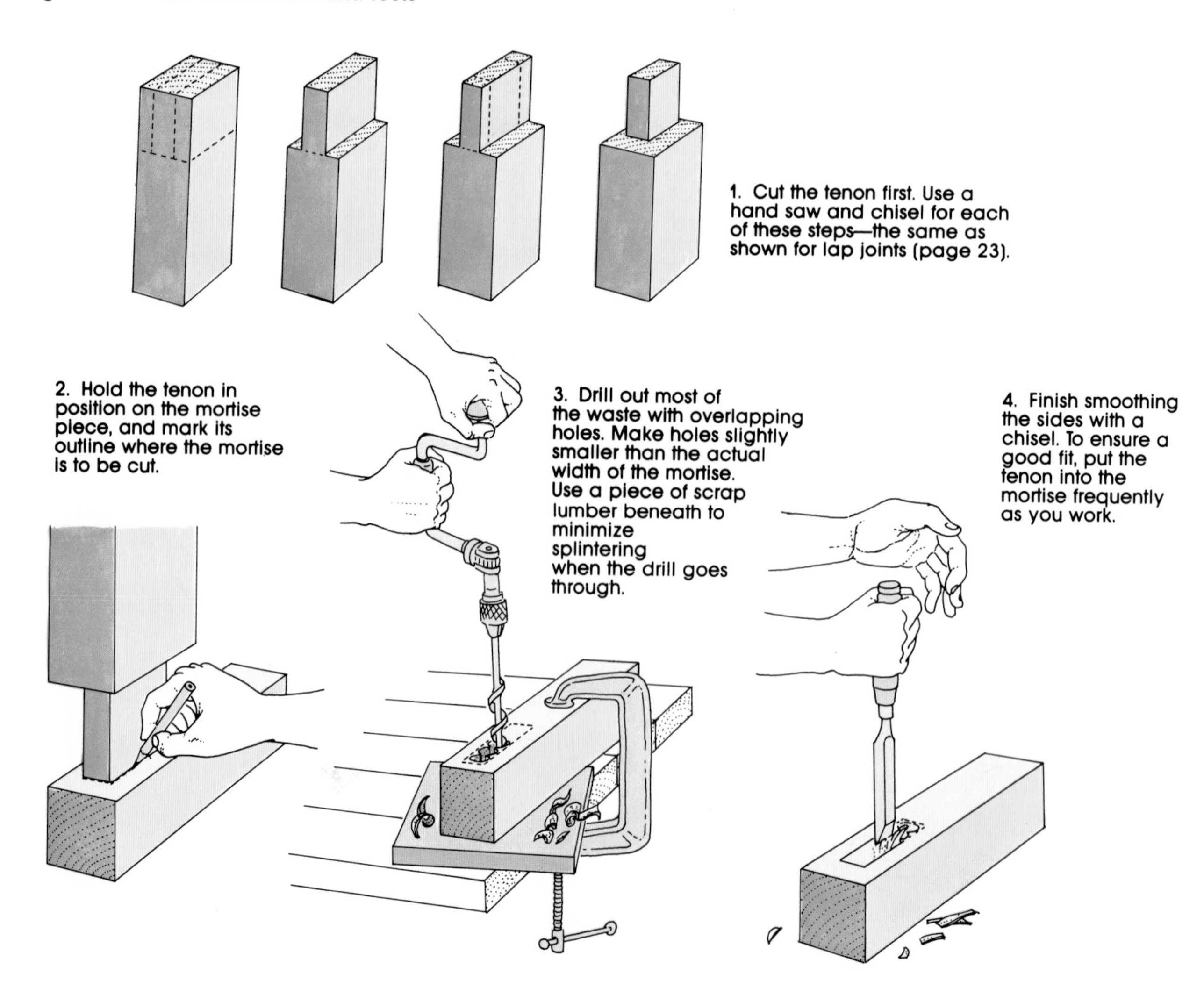

1. Cut the tenon first. Use a hand saw and chisel for each of these steps—the same as shown for lap joints (page 23).

2. Hold the tenon in position on the mortise piece, and mark its outline where the mortise is to be cut.

3. Drill out most of the waste with overlapping holes. Make holes slightly smaller than the actual width of the mortise. Use a piece of scrap lumber beneath to minimize splintering when the drill goes through.

4. Finish smoothing the sides with a chisel. To ensure a good fit, put the tenon into the mortise frequently as you work.

# HANDSAWS

## Types of Handsaws

Although there are more variations in handsaws than in any other tools, you can select a good saw if you have some basic information. Every shop must have a good crosscut saw. It is designed to cut across the grain, as opposed to the ripsaw, which is designed to cut with the grain. Other important saws include the compass saw, backsaw, coping saw, hacksaw, and for occasional detail work, the dovetail saw.

### Crosscut Saw

A crosscut saw should have a highly flexible steel blade that gives a clear ring when tapped. It should be thinner along the back than along the cutting edge to lessen the chances of binding (getting pinched in the wood). Finally, a good crosscut blade is tapered from the narrow toe to the wider heel.

The number of cutting teeth per inch is another critical factor in the selection process. Crosscuts generally have from 7 to 12 teeth per inch. For general shop and construction work, 7 or 8 teeth per inch gives efficient but slightly rough cuts. For fine, smooth cuts, select a saw with 10 to 12 teeth per inch. This information is stamped on the blade of the saw near the handle.

The teeth on a crosscut saw are set out on alternate sides about one-fourth of the blade thickness. The saw cuts with a knife-like or slicing action and results in a cut that is slightly wider than the blade, which helps prevent binding.

### Ripsaw

The ripsaw is normally 26 inches long and has heavy, chisellike teeth. There are 5½ teeth per inch, and they are designed to cut with a chisellike action through long wood fibers that run with the grain. As with a crosscut, a good ripsaw is tapered from the back to the teeth and from the toe to the heel.

### Compass Saw

The compass saw has a long, flexible blade designed to cut curves in wood. It is particularly good for working in tight corners or cutting out from the middle of a piece of wood. In such a case, a hole is drilled in the wood and then a compass saw is used to finish the job. The compass saw is sometimes called a keyhole saw, but this is actually a smaller, finer saw that was once used to cut keyholes in doors.

### Backsaw

This is a short saw with a rigid back designed to prevent it from flexing, which could throw it off the angle of the cut. It usually has 12 teeth per inch for smooth cuts. The backsaw is commonly used with a miter box to cut molding and trim at precise angles for a smooth fit.

### Coping Saw

The coping saw has a thin blade with up to 20 teeth per inch, and is designed for cutting circles and patterns with tight curves. The teeth on the blade are all angled in one direction, which affects your cutting. For general woodwork, the teeth should be pointed away from you so that they cut as you push the saw. For finer work and more control, the teeth should be pointed toward you, cutting as you pull the saw.

### Hacksaw

Although technically not a woodworking saw, the hacksaw is vital to any shop or construction project. There are several varieties of hacksaw blades. The coarser the teeth, the faster you can make the cut. Generally, blades with 14 teeth per inch should be used to cut aluminum, brass, copper, bronze, or steel. Blades with 24 teeth per inch should be used for iron, steel, iron pipe, or copper or brass tubing. The teeth should point away from you, since the cut is made on the push stroke.

Another handy tool is the mini-hacksaw. This saw can be used in many places the conventional hacksaw cannot, such as for cutting nails under shingles.

### Dovetail Saw

The dovetail saw is like the backsaw, only smaller, and is designed for fine cuts made by cabinetmakers who delight in using hand tools.

**Types of Hand Saws**

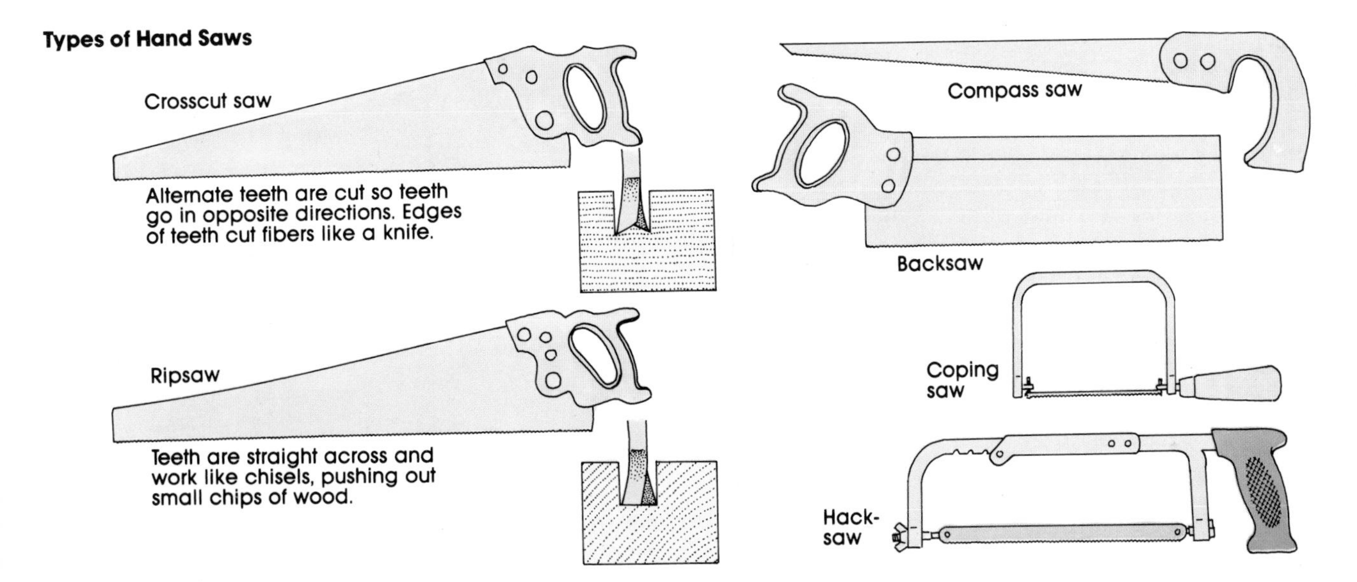

## Buying a Handsaw

A good handsaw, with care and periodic sharpening, is a tool that will serve you well for years. It is therefore wise to buy the best. One of the first things to look for in saw quality is the blade itself. It should be made of tempered spring steel. This information may or may not be stamped on the blade, but you can tell by means of two simple tests. First, hold the saw by the handle and snap the blade sharply with your finger. There should be a clear ringing sound, not a dull thump. The second factor that goes with tempered steel is springiness. Hold the handle firmly with one hand and then grasp the toe with the other hand. Bend the saw in a half circle and release it. It should snap right back in place.

A good saw has a blade that is tapered so the widest part is near the teeth and the narrowest part is near the back. This design means there is less drag on the saw as you cut deeper into the wood. Similarly, a good saw is narrower at the toe than the heel so the toe will not easily catch in the saw kerf (the slit cut by the saw) and force the saw to bend.

The teeth on a tempered steel blade will take and hold a good edge for a long time. Sharpening a saw is best done by the professional at a saw shop.

The handle should be a smooth hardwood that provides a comfortable handhold.

**Sawing Technique**

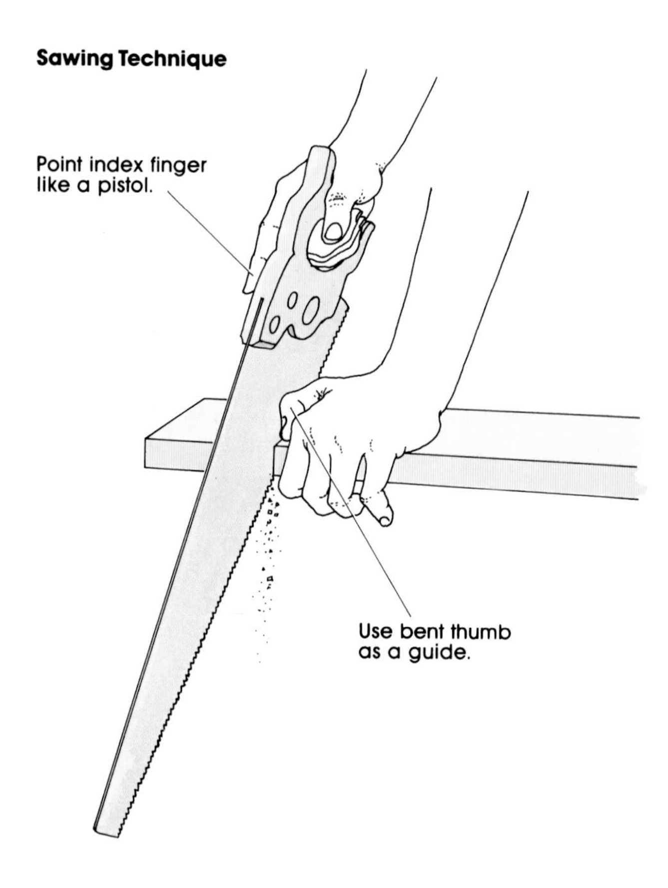


## Sawing Techniques

Learning to saw a straight line is not as simple as it may look. It takes practice and concentration.

One of the first rules to remember is *always save the line*. If you have measured accurately, the line you draw should be precisely at the end of the piece you want. If you cut *on* the line, the cut (or more properly, the kerf), which is about ⅛ inch wide, will leave your board ⅛ inch short.

## Miter Box

The backsaw is often used with a miter box, which is a tool designed to cut precise angles or miters. It is particularly useful when cutting angles for picture frames or molding.

The simplest miter box, normally available in hardware stores, is made from hardwood. With it, you are limited to cutting either a 90 or 45 degree angle because that is all it's designed for. The wooden miter box, by its construction, also limits you to cutting fairly narrow pieces of wood (see illustration).

The steel miter box has an open shelf on which you can place much larger pieces to be cut. The backsaw is held by two precision guides that are calibrated to cut any angle you set them at between 30 and 90 degrees.

In order to use either the wooden or the steel miter box, place the piece to be cut in position under the saw at the chosen angle. Hold the wood firmly with one hand, lower the saw, and then cut with gentle strokes so you won't jar the wood out of position. Remember that the line you have drawn on the wood represents the end of that piece, so cut just along the *waste edge* of the line, not along the line itself. If you cut directly on the line, you will shorten the piece by the thickness of the saw.

**Wooden Miter Box**

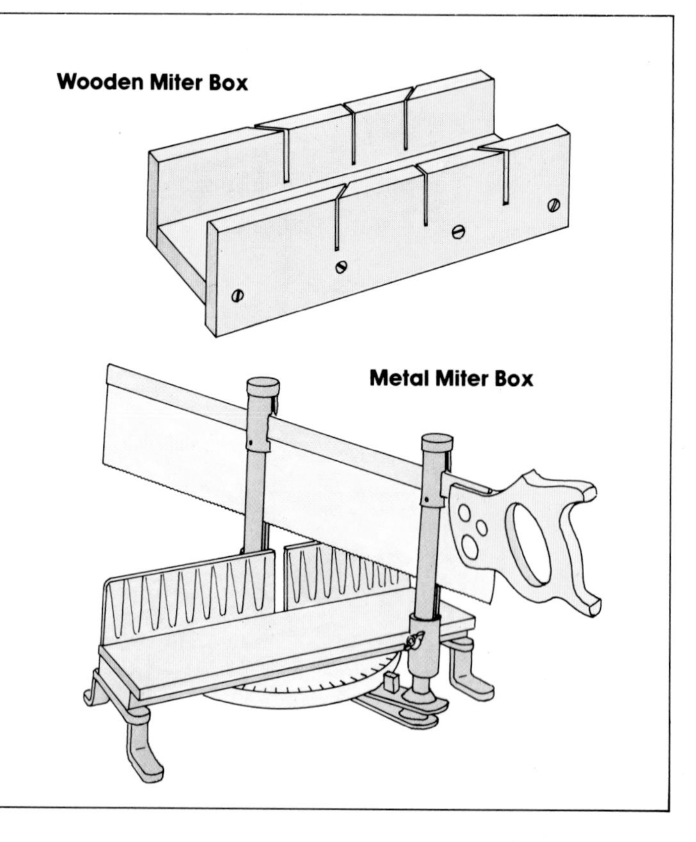

**Metal Miter Box**

So, when you saw a board, first make sure it is braced and steady. Then, using your bent thumb as a guide, place the saw on the waste edge of the line. Draw the saw slowly toward you to start the kerf. Repeat this process a second time, with your hand moved clear of the saw. Then, holding the saw at about a 45 degree angle, begin the push–pull process of sawing. Keep your index finger outside the handhold. Point it like a pistol and you will find it helps to guide the saw in a straight cut. Apply only a light downward pressure on the saw. Too much pressure can cause the saw to bind and spring out of the kerf, often landing on your fingers.

The saw blade must not lean left or right as you cut. Keep it vertical. If you have doubts at the beginning, use a try square (see page 59) on the board to check that the blade is vertical. Keep your eye over the saw back and on the line.

If you get off the line as you cut, bend the saw slightly to bring it back in line. This, however, will generally cause your saw to bind to some degree, making your job much more difficult. If you find that you can't work the saw anymore, start over from the other side.

When you near the end of your cut, brace the board with one knee to keep it from moving, and with your free hand steady the waste piece about to fall off. Make the last few cuts short and rapid, and support the waste piece so it won't split away and splinter the good piece.

Sometimes you have to saw off a very small piece—¼ inch or less—from the end of a board. The problem is that the narrow end piece will keep breaking away, leaving nothing to support the saw. You can counter this by sawing through two boards at once. Place the board to be trimmed underneath.

### Binding

Binding, or getting the saw stuck in the wood, is a common problem. One of the chief reasons for binding is that the board you are cutting sags in the middle and forces the kerf closed, pinching the saw. Let the waste end hang free, pulling the kerf slightly open. Another source of binding problems is an improper set (not enough) on saw teeth. If you have had your saw checked for a proper set at your saw shop, and have followed all the other precautions noted here, have your saw checked by a professional. Green wood and pitchy wood can also cause binding. When binding occurs in green (freshly cut) wood, keep the saw lightly oiled. When it occurs in pitchy (resinous) wood, which gums up the saw, coat the saw with turpentine or kerosene.

When you work with a ripsaw on a long board, sometimes the kerf will spring back together and bind the saw. Counter this by jamming a screwdriver into the kerf behind the saw to hold the kerf open.

When you rip long boards or cut pieces of plywood, you will need to support both sides of the wood. One piece hanging down too far will tend to bind the saw and may also break off before you finish the cut. If the piece is too long for you to support with one hand while you saw, you must support it with a sawhorse or chair or whatever you can find.

### Circular Saw

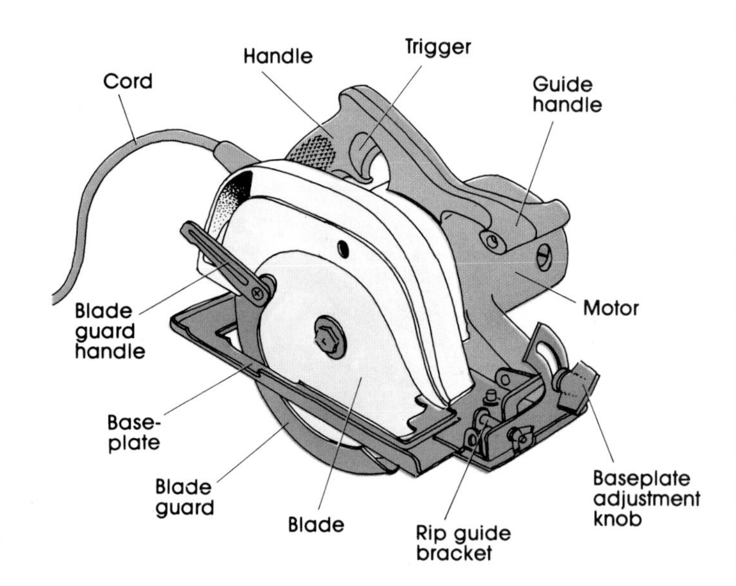


The biggest time saver in a carpenter's array of tools is the portable circular saw. It is a must for the most basic workshop because of its time-saving qualities and versatility. A circular saw will cut any board in a fraction of the time it takes to cut by hand. Within the limitations of its blade diameter, it can be adjusted for either shallow or deep cutting. By adjusting the baseplate on the saw (see illustration), you can cut bevels of any angle between 90 and 45 degrees. The saw can also be fitted with a variety of special blades for crosscuts, rips, a combination of both, or for cutting plywood, masonry, or light metal and plastics.

Circular saws are made with either metal or hardened plastic housings. The on–off switch is a trigger built into the handle. All circular saws have a retractable blade guard. When the saw is not in use, the spring-activated guard cups the blade. When the saw is being used, the guard slides up and out of the way as the blade enters the wood.

In addition to the adjustable baseplate that controls bevels and saw depth, good saws have an adjustable rip fence that fits along the edge of the board to be cut and helps guide the saw.

### Buying a Circular Saw

Circular saws are designated by the size of the largest blade they will accept. These sizes normally range from 6½ to 10 inches, with the standard homeowner's saw being 7¼ inches. The smaller saws are for trim and flooring work, and the larger ones are for professional construction carpenters. The 7¼-inch saw blade has enough cutting depth to handle most of your shop and construction projects. It will also cut cleanly through a 2 by 4 when set at a 45 degree angle, whereas a 6½-inch blade will just barely make it.

Most 7¼-inch saws are likely to be 1¼, 2, or 2½ hp. The more horsepower, the smoother the saw runs and the faster it cuts under a load. The less horsepower, the faster the motor will wear out under use, so choose the highest horsepower you can afford.

### Circular Saw Blades

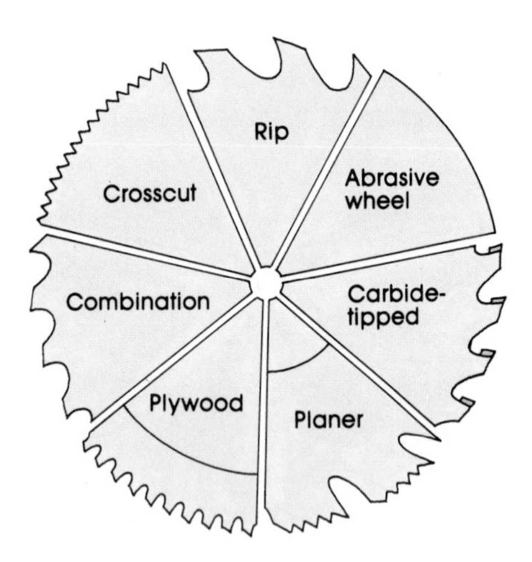


Just as important as the saw itself are the blades you use in it. The blades, which vary in number, style, and hardness of the saw teeth, are designed for specific jobs. Don't, for instance, try to cut fiberglass with a plywood blade. It just wasn't designed for that material.

Saw blades are sold in individual packages that not only tell you what that blade is for but also list and illustrate other blade styles. The descriptions that follow will help you select the right blade for the right job.

*Crosscut blade.* This blade has a series of evenly spaced, medium-sized teeth that are set (bent) alternately to the left and right. Each tooth is sharpened on the inside of the tooth set to provide a smooth cutting action across the grain.

*Rip blade.* The teeth on a rip blade are also set alternately to the left and right, but unlike crosscut teeth, they are sharpened on the top, not on the inside. They are, in effect, a series of chisels that scoop out wood as the saw moves with the grain.

*Combination blade.* This blade, which incorporates features of both the rip and crosscut blades, is normally sold with new circular saws. It is the best choice for general shop and construction work because it does both jobs about equally well.

*Plywood blade.* With its many small teeth, this blade gives clean cuts through plywood. Crosscut and combination blades can also be used in cutting plywood, but the larger teeth will cause the plywood to splinter badly. Rip blades will really tear up plywood and should not be used for that purpose.

*Carbide-tipped blade.* This blade handles general-purpose jobs like the combination blade, but the cutting edges are made from extra hard carbide steel. This blade is a good choice on construction sites where you may be cutting wood of different hardnesses.

*Abrasive wheel.* These wheels have no teeth but literally grind their way through the material being cut. Some abrasive wheels are made for cutting masonry or fiberglass, others for cutting light metal. Don't try to cut something the wheel is not specifically designed to cut. While the blades are tough, they are also brittle. Never twist the blade in a kerf with the saw running; it could shatter, with disastrous results to you and any bystanders.

## Using the Circular Saw

Although it is a powerful and versatile tool, the circular saw is no better than the person using it. Only with practice and experience will you realize how useful it is.

Remember that the circular saw cuts *up,* which can cause splintering on the top of the piece being cut. Therefore, always keep the best face of the wood down. This is particularly true with plywood.

The basic techniques for making a variety of cuts are described below.

**Crosscut.** Trimming boards to length is the staple diet of the circular saw. Place the saw in position with the baseplate flat. Start the saw before the blade touches the wood. Move across the wood smoothly and evenly. If the blade begins to slow or bind, keep the motor running while you back the saw out a little, then move ahead again, this time more slowly. Remember to keep the saw blade just on the waste side of your cutting line. Most circular saws have a notch on the baseplate that is supposed to serve as a "sight" on the cutting line. For accuracy, never rely on that; always watch the saw blade and the line.

**Rip cut.** When you cut down the length of a board, extend the rip fence on your saw to the edge of the board and

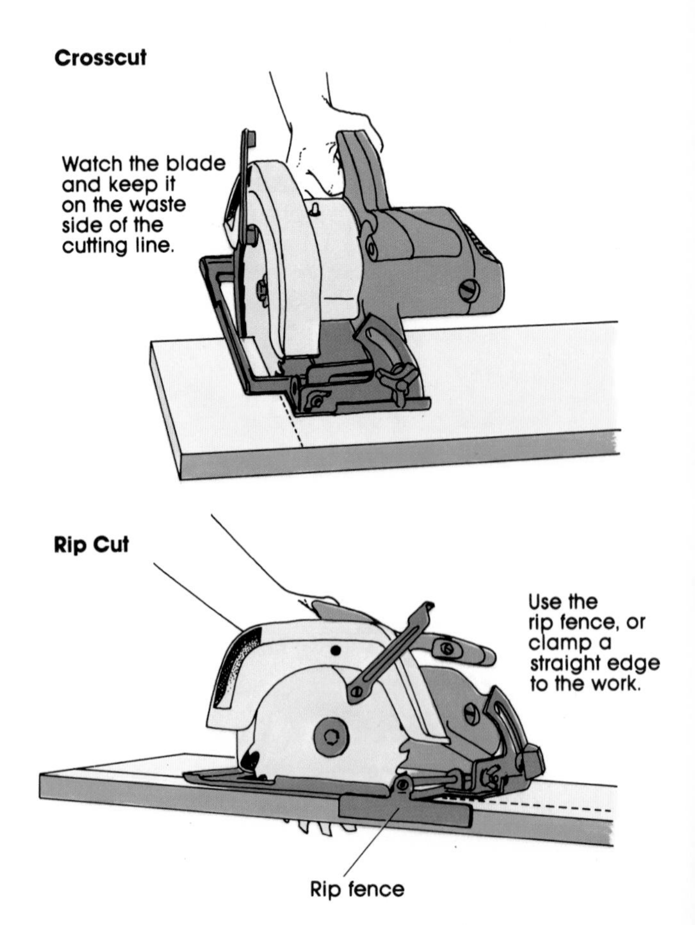

let it help guide the saw straight. If the board is too wide to use the rip fence, clamp a straight edge on the board next to the saw to guide it. This is particularly useful when you cut plywood, which is difficult to cut straight because of its relative thinness. If the rip cut tends to bind, jam a screwdriver or chisel in the kerf to spread it away from the blade.

**Bevel cut.** A guide on the baseplate is calibrated from 90 to 45 degrees. To make bevel cuts, loosen the adjustment knob holding the baseplate and tilt the base to the desired angle. To make a smooth and square bevel cut, start the cut on the line, then clamp a board to the work next to the saw to guide it straight across.

**Pocket cut.** It is often necessary to start a cut in the middle of a board. To do this, you make the pocket cut. Pull the blade guard up and align the saw blade directly over the line to be cut. Keep the blade free of the wood before turning on the motor, then slowly lower the blade into the wood. Do not start cutting forward along the line until the baseplate is firmly seated on the wood.

**Dado and rabbet cut.** These two cuts (described on pages 24 and 25) can be made quickly with a circular saw. First, by adjusting the baseplate, set the saw blade to the required depth. To make a dado, first cut the outer edges, then make several passes through the center portion to remove most of the waste. Trim up what is left with a chisel.

To make the rabbet, place the piece to be cut in a vise, butt end up, and make the vertical cut first. Next, reset the blade to the required depth and cut across the face of the board to complete the rabbet.

## Safety

The following basic safety precautions are essential when you are using a circular saw:

- Never remove or wedge up the retractable blade guard. It is designed not only to save fingers but also to allow you to put the saw down while the blade is still spinning.
- When you cut in tight, high places, keep your face away from the back of the saw. It can jam and kick back.
- Never work on the saw while it is plugged in.
- Don't leave the saw plugged in around the work site where a child might pick it up and turn it on.
- Always wear safety glasses when using a circular saw and especially when using the abrasive wheel.

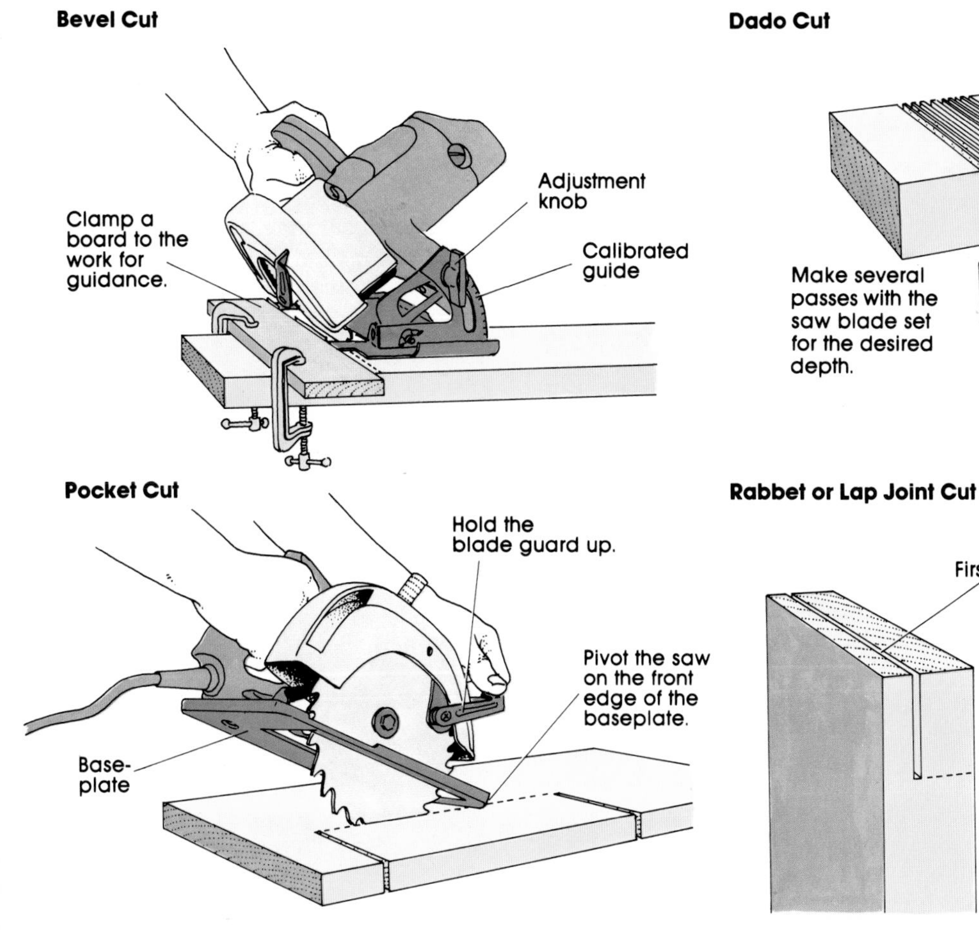


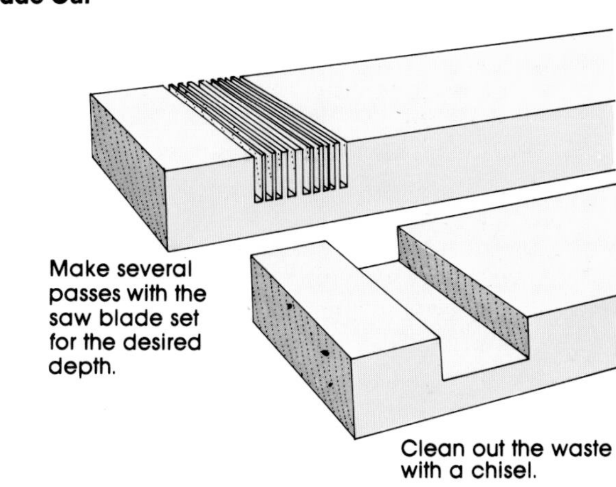


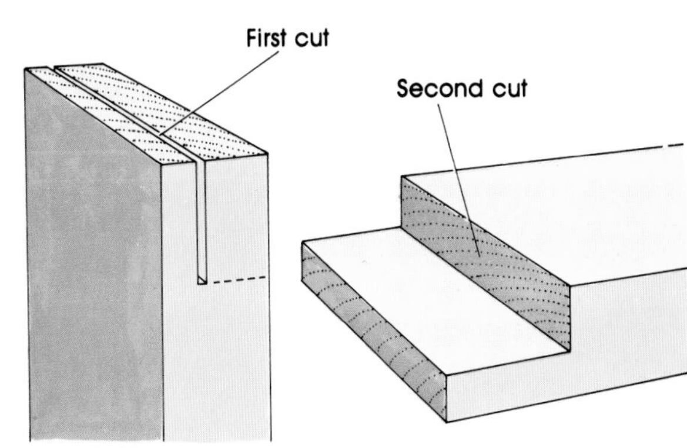

# SABER & RECIPROCATING SAWS

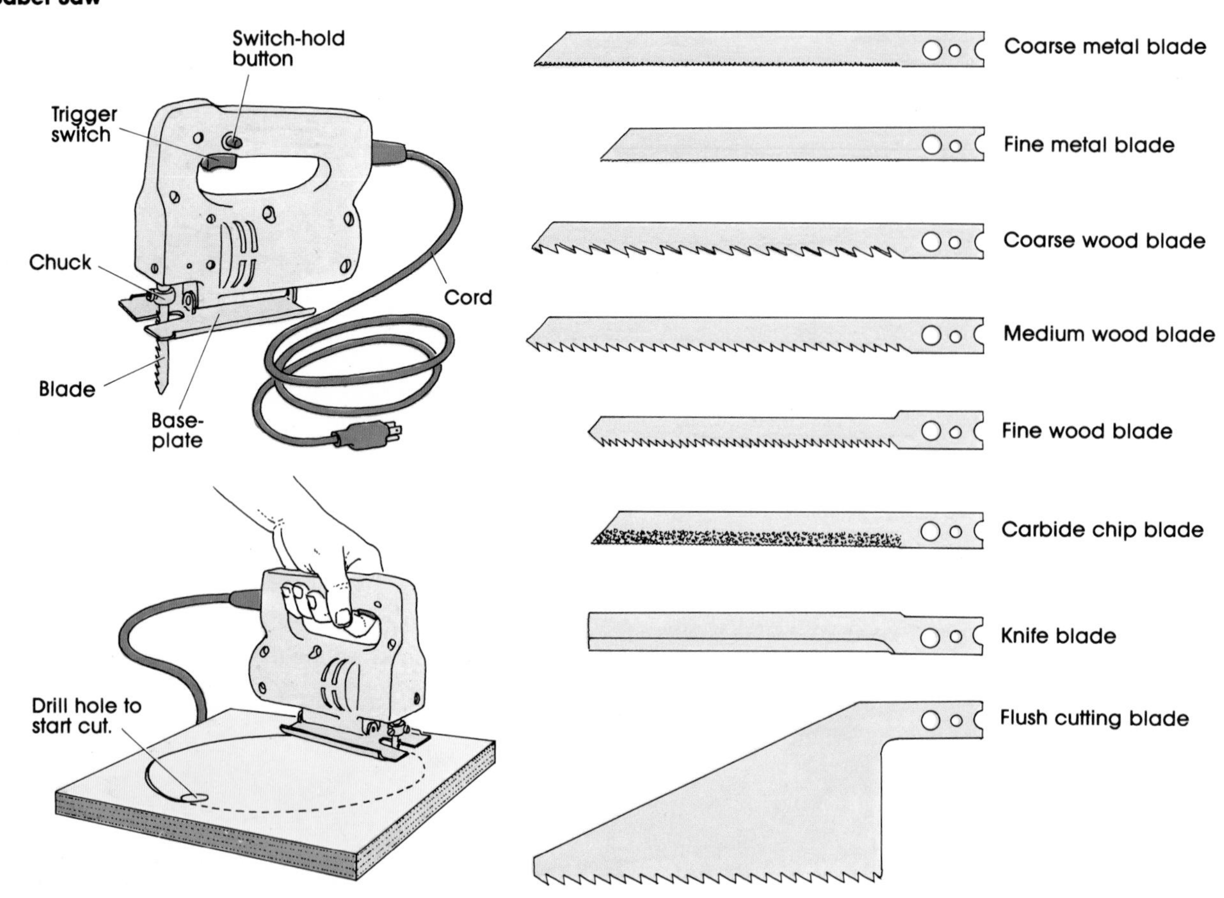


The saber saw or jigsaw is primarily used for cutting curves. It is a valuable tool for any workshop because it cuts efficiently with a wide variety of blades, each designed for a specific task. Light enough to operate easily with one hand, it also comes in handy for tight places where a circular saw won't fit.

## Buying a Saber Saw

Saber saws range in horsepower from a low of ⅙ hp to ½ hp or more. The more horsepower, the stronger the sawing action and the least wear on the motor. For general shop use, ½ hp is sufficient; saws rated above that are for commercial use.

One-speed, two-speed, three-speed, or variable-speed motors are available. You should select two-speed or more because different speeds are used to cut different materials. The harder the material, the slower the cutting speed should be. If you try to cut hard material such as metal at high speed, the blade will overheat and break.

In general, use high speeds for clean cuts in wood, plastic, or fiberboard; medium speeds for soft metals such as aluminum or copper; and slow speeds on harder metals such as bronze or steel.

A good saber saw will have a baseplate that tilts from 90 to 45 degrees for beveled cuts.

### Saber Saw Blades

The blade you choose should be determined largely by the material you are cutting. The fewer teeth per inch in the blade, the coarser the cut. For making a finish cut or for cutting hardwood, use a blade with 14 teeth or more; for rough cuts in wood, use a blade with fewer but larger teeth. If you will be cutting metal, choose a blade made from hardened steel that will not easily overheat and snap. The chart on this page provides a quick reference.

## Using the Saber Saw

Try not to go too fast when using this saw. Pushing the saw forward too hard will not only cause the blade to overheat and snap, but will also soon wear out the motor.

To cut a circle, first drill a hole in the workpiece, then insert the blade and begin the cut. Always use the narrowest blade possible for tight circles or curves so the kerf will not restrict the turning abilities of the blade. Never force a blade around a curve or it may snap. If necessary, back up and make several passes through the waste portion to complete the curve.

To make a pocket cut (a cut through the middle of a board), rock the saw forward so the blade is clear of the wood, position it over the line, start the motor, and slowly lower the blade into the wood. Start cutting forward only when the baseplate is seated on the wood.

The baseplate can be tilted from 90 to 45 degrees for making miter cuts. To make them accurately, first position the blade without quite touching the wood, then place a square on the board next to the saw to guide it as you begin the cut.

## Reciprocating Saw

This big brother to the saber saw is commonly found on construction sites because of its great versatility. Carpenters use it to rough cut big beams or posts; plumbers use it to cut holes in floors for piping; and heating contractors use it to make room for duct work.

The reciprocating saw is very useful to have around if you are remodeling a house. For example, you can use it to cut out a portion of a wall for a new window or door. Before attempting anything like this, however, make sure that any wiring in the wall has been removed. You should also be extremely careful *not* to cut through any wiring, or you'll get a shock.

On many job sites the reciprocating saw is referred to by a popular brand name, Sawzall. It is available with one, two, or variable speeds. Choose a variable-speed saw if you will be cutting much metal in your projects; it will operate at a speed slow enough to keep the blade from overheating and breaking.

Blades for reciprocating saws range from 2½ to 18 inches in length and can be mounted with the teeth facing either up or down on the saw, depending on your needs. The better the saw, the more blades will be available to you, including a wide variety for both wood- and metal-working. The following blades might be found in a basic set: 6-inch smooth cutting, 6-inch flush cutting, 5-inch metal cutting, and 4½-inch sheet metal cutting.

**Reciprocating Saw**

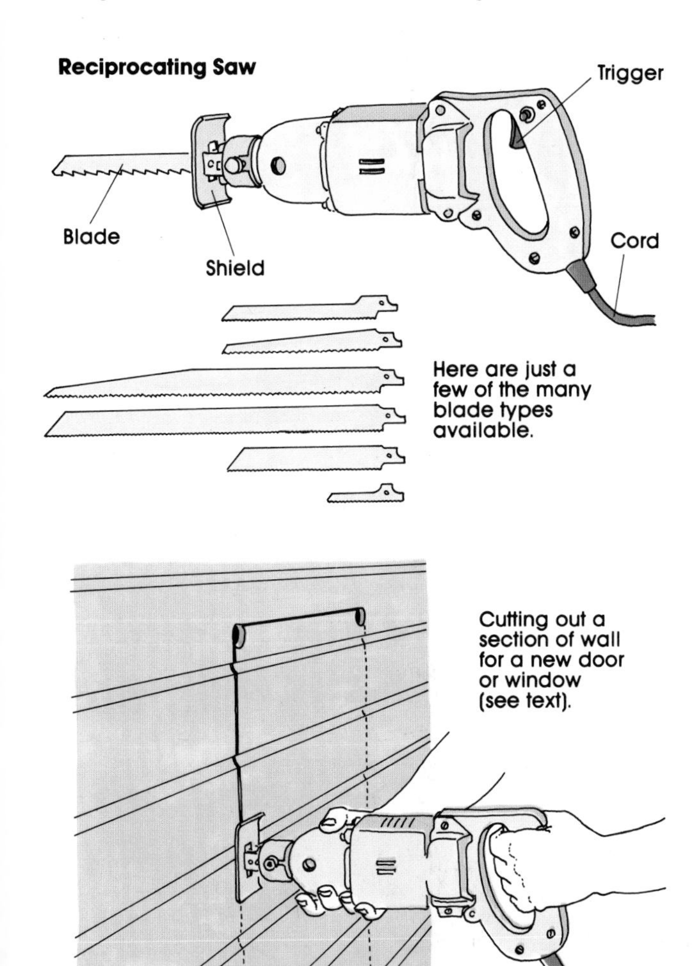


Here are just a few of the many blade types available.

Cutting out a section of wall for a new door or window (see text).

**Parts of a Table Saw**

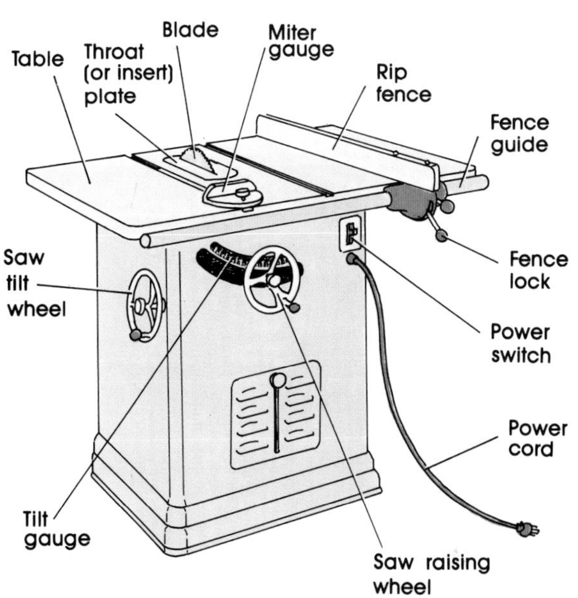


The blade on a table saw can be raised or lowered to adjust the cutting depth, and it can be tilted up to 45 degrees for making bevel cuts. Unlike the radial arm saw, which fits snugly against one wall of a shop (see page 37), the table saw must be placed in the center of the floor so you can work around it. Contractors generally prefer the table saw because it can cut large sheets of plywood.

The table can be made from aluminum, steel, or cast iron. Because of its light weight, an aluminum table may have to be bolted to the shop floor to keep it from rocking when you are working with it. But the light weight could be an advantage if you must often move the saw to different work sites. The additional weight and stability of steel or cast iron is usually preferable for general shop use.

The size of a table saw is designated by the maximum size blade it can use. Blades range from 6½ to 12 inches in diameter. The smallest size would be used by a cabinetmaker cutting relatively thin pieces of wood, while the 12-inch saw is generally for professional contractors. The 9- or 10-inch saw is a popular size for the home. A 9-inch saw will just barely cut through a 2 by 4 at a 45 degree angle, while a 10-inch saw will make the same cut quite easily.

Two devices enable you to make consistently accurate cuts with the table saw: the fence and the miter gauge. The *fence* is a flat metal bar, usually with a scale on it, that remains parallel to the saw blade. It can be locked in place any distance from the blade and controls the width to be cut.

The *miter gauge* moves in a groove parallel to the saw blade, but the end that rests against the wood is adjustable. Normally it is kept at 90 degrees to provide a square cut when trimming wood to length, but it can be adjusted to any angle up to 45 degrees.

The blade itself can be adjusted to cut any bevel up to 45 degrees. It also adjusts up and down to a variety of heights.

## Buying a Table Saw

When you shop for a table saw, one of the first considerations should be the blade size (discussed on page 33). A 9-inch blade is the minimum size necessary for general home use.

The table itself should be well braced and large enough for the work you will be doing on it. The smaller it is, the more problems you will have keeping large boards or plywood steady while cutting them. Some tables have extensions on the side for added support.

Two safety features are essential for any table saw. The first is a removable saw guard, usually plastic, that allows you to observe the cut being made. It is angled up slightly in the front to ride smoothly over the wood being run through the saw. Although it must be removed when working on large pieces of wood, you should make a habit of leaving it in place. A moving blade is nearly invisible, and accidents do happen.

The second safety feature is a splitter, a device that rides either directly under or over the saw and keeps the kerf open. This helps prevent binding, which can hurl a piece of wood back at the operator. The splitter should also have anti-kickback fingers, which normally ride with the wood but will sink into it if it starts to kick back.

You should also consider the following questions before buying a table saw:

- Is the fence bar firmly in place or does it rattle a little?
- Can you lock both the front and the rear of the fence with a single handle in front?
- Does it have a large insert plate to make changing blades easier?
- Does it have enough horsepower for your needs? Ask the salesperson about the actual horsepower, not what the saw develops. Sometimes twice the actual horsepower is developed just for the initial surge. How much horsepower you need depends on what you are willing to pay. Remember that the smaller the horsepower, the harder the motor has to work and the sooner it will wear out.
- What actually comes with the saw? Some are sold without the motor, for instance. Don't be afraid to bargain with the salesperson for a few extras.

### Table Saw Blades

There are two basic types of blades for both the table saw and the radial arm saw: the set tooth blade and the hollow ground blade. Each type is available with different styles of blade teeth designed for specific cuts, such as crosscut, rip, or plywood (see illustration).

The *set tooth blade* makes a cut, or kerf, that is wider than the blade because the teeth are set (bent) alternately to the left and right. These blades are recommended for general shop use.

The teeth on the *hollow ground blade* are not set to make a wider kerf. Instead, the blade itself is ground so that the teeth are the widest part of the blade. The blade becomes thinner toward the arbor hole in the center. This type of blade, which can be identified by the unpolished metal around the arbor hole, gives clean cuts but tends to bind more easily than the set tooth blade.

The *combination blade* is the first choice for general use. It has alternating teeth designed for both ripping and crosscutting.

The *carbide-tipped blade* has teeth tipped with carbide steel and will long outlast ordinary blades. It is often used as a combination blade.

The *ripping blade* is worth buying if you have a lot of ripping to do. It will do the job better than a combination blade.

The *crosscut blade* gives a very smooth finish to the cut and is excellent for cabinetwork.

The *plywood blade* has small, sharp teeth designed to cut plywood without splintering.

A *standard dado blade* has two outside cutting blades with chippers of varying widths sandwiched between them. Another excellent alternative is the *adjustable dado blade,* which is calibrated so that it can be set to wobble up to ½ inch to cut the desired width.

### Changing Blades

This is the most common maintenance task on the table saw, but the first time can be confusing. First, unplug the machine so there won't be any chance of it starting. Lift out the throat (or insert) plate around the blade and then lower the saw all the way. Use a box end wrench so you won't risk slipping and damaging the nut. This is a left-hand thread, so you loosen it by turning it clockwise. If necessary, block the blade with a piece of scrap wood to keep it from turning.

Set tooth blade
Cross section
View of teeth
Hollow ground blade
Cross section
View of teeth
Plywood blade
Carbide-tipped blade
Ripping blade
Crosscut blade
Combination blade

**Dado Blades**

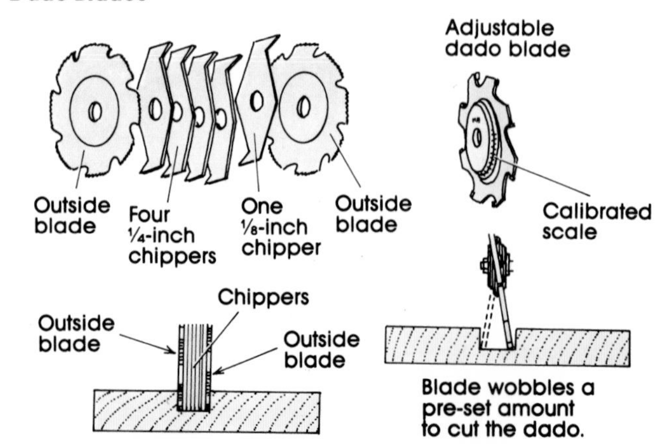

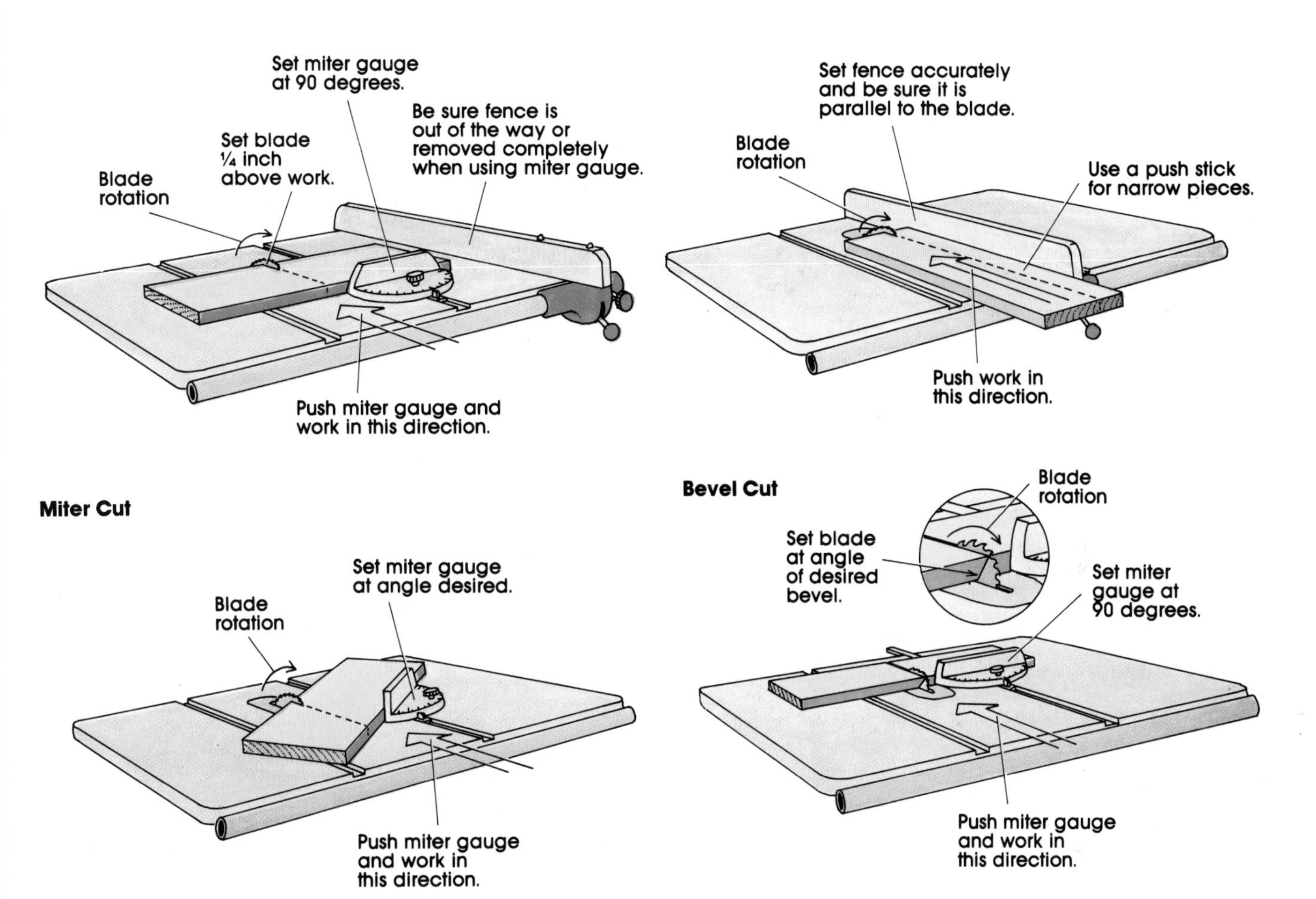


## Basic Cutting Techniques

**Crosscut.** Before using the saw, check that the blade is perpendicular to the table by placing a try square (see page 59) on the table and sliding it up to the blade. Make any adjustments necessary to align the blade. In order to make a crosscut, hold the board against the miter gauge and slide it across the table. Hold the wood only on the side against the miter gauge—never on both sides of the cut. Holding the wood on both sides may cause binding, which will jerk the wood and possibly throw your hand into the saw blade.

**Rip cut.** To rip a board accurately, make sure the fence is perfectly square with the blade. Check this by measuring the space between the blade and the fence at the front and rear of the blade.

When you push through short or narrow pieces, never let your fingers come closer than the width of your hand to the blade. Make a push-stick by notching the end of a piece of 1-by-2 scrap. When you work with a board too long to handle easily, rip it for half its length, then shut off the saw. Reverse the board end for end, carefully align your saw cuts, and complete the cut.

**Miter cut.** First, set the gauge on the miter to the desired angle. Hold the board firmly against the miter and then slide it forward across the table to make the cut. When using a metal miter gauge, you will notice that the wood tends to slip slightly toward the saw. You can correct this by screwing a piece of wood to the gauge and letting the tips of the screws protrude slightly on the face of the wood. These sharp points will anchor the piece you are working with.

**Bevel cut.** To make the straight bevel cut, tilt the saw blade to the desired angle by rotating the bevel wheel. The wheel has calibrations on it that range from 90 to 45 degrees.

With the saw blade tilted to the desired angle, hold the stock firmly against the sliding miter gauge and push the gauge and wood together across the table to make the cut.

**Bevel miter cut.** To make this cut, first set the saw blade to the desired bevel angle. Next, adjust the miter gauge. Place the stock to be cut on the gauge and, before turning on the saw, align the saw blade with the mark to be cut. Remember that the cut should be made just along the waste edge of the line and not on the line itself. Cutting the line will shorten the board by about the thickness of the blade.

**Bevel rip cut.** The bevel angle can be determined either by tilting the saw blade to a predetermined angle or by aligning the blade with a bevel angle marked on the butt end of the stock. Once the bevel angle is determined, slide the fence up the board, check that board and blade are properly aligned over the cut mark, then push the board through.

**Dadoes, rabbets, and grooves.** When made with a blade, these cuts are all executed in the same manner. First, adjust the dado blade to the desired width of the cut. Methods of adjustment differ slightly according to the make, but all come with complete instructions.

Next, set the blade to cut at the desired depth. To do this, place the butt end of the board to be cut next to the blade, then raise or lower the blade to the proper setting. This should be marked on the butt. If you need dadoes, rabbets, or grooves wider than the dado blade, make repeated passes over the blade with the stock to be cut.

Dadoes, rabbets, and grooves are discussed in more detail on pages 24 and 25.

**Tenons.** A dado blade can cut tenons very efficiently. Mark the board to be cut in the manner described on page 25. Set the blade to the required depth and trim one side of the tenon by making several passes over the dado blade. When one side is cut, turn the board over and do the other side in the same manner. If the two edges of the tenon need to be trimmed, you should use a backsaw (see page 27).

## Table Saw Safety Rules

- Always wear safety goggles.
- Never reach across the saw blade.
- Never stand directly in line with the saw blade.
- Keep the saw guard, splitter, and anti-kickback fingers in place whenever possible.
- Never let your fingers come closer than your hand's width to the blade. Use a push-stick for small or narrow pieces.
- Keep the wood firmly against the miter gauge or the fence. Never push the wood through freehand.
- Always turn off the saw before trying to free a jammed blade.
- Never use the fence as a stop guide when crosscutting. It can cause a dangerous bind.
- Don't hold both ends of the wood when crosscutting.
- Don't work around the saw when wearing loose clothing, such as dangling sleeves or a swinging necktie.
- Keep the top of the blade only 1/4 inch above the stock.

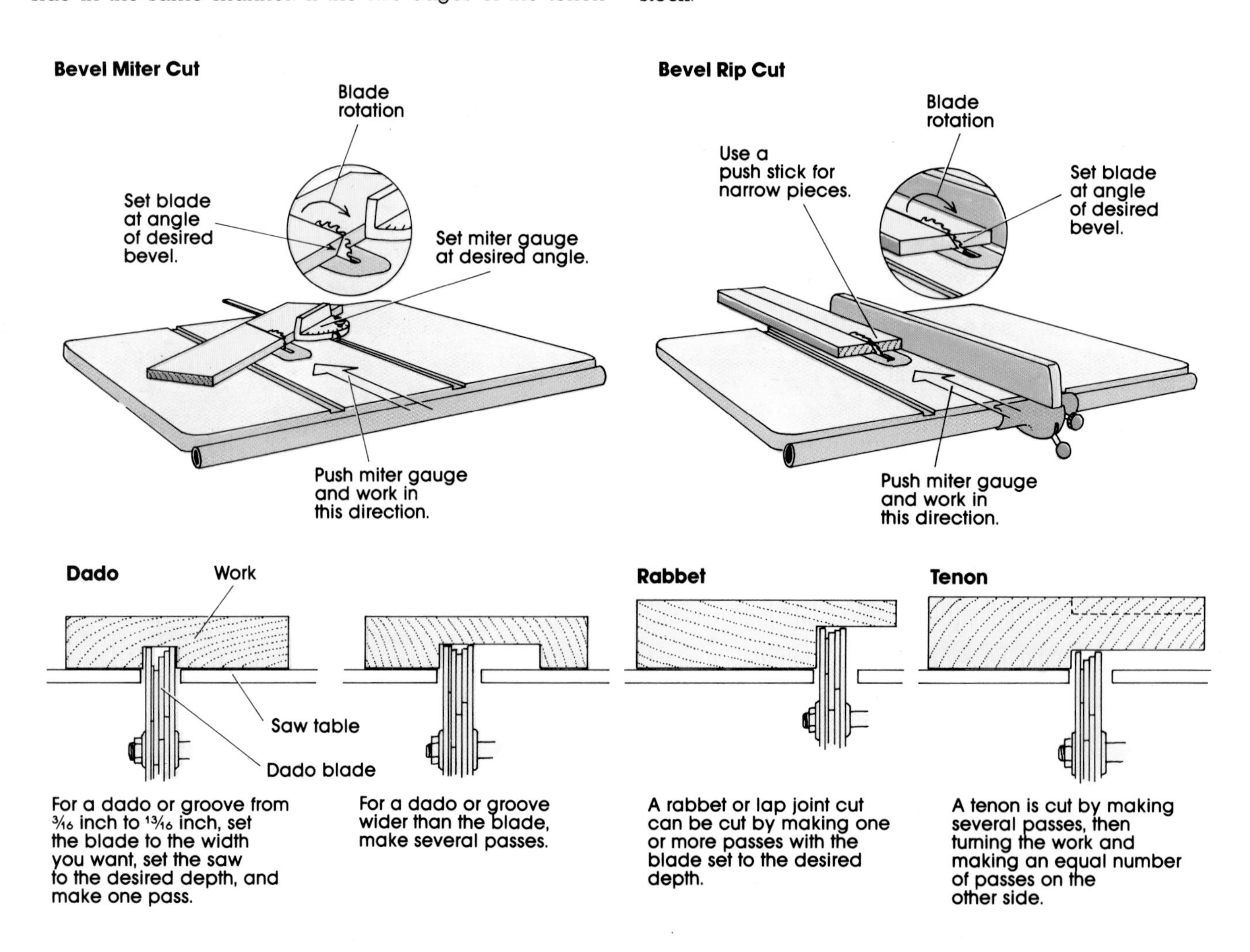

# RADIAL ARM SAW

**Parts of a Radial Arm Saw**

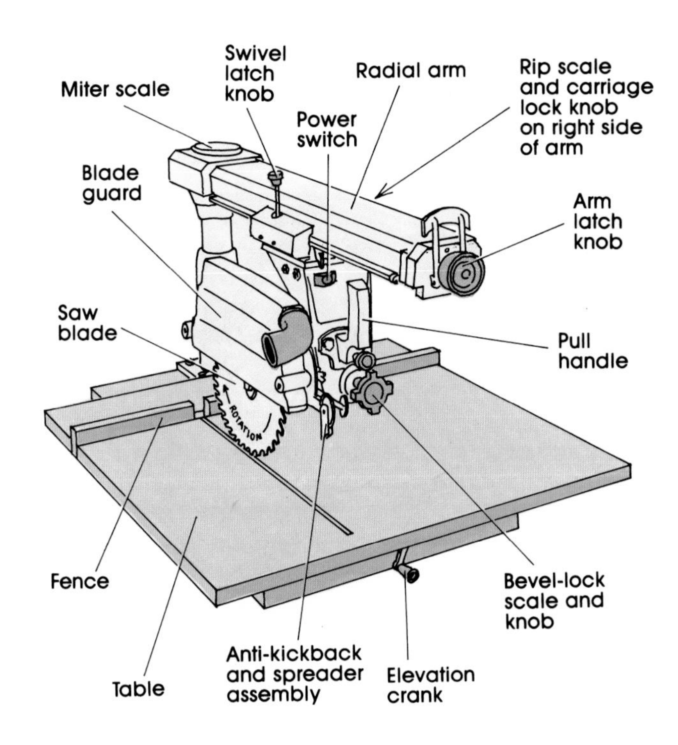


The radial arm saw is one of the most versatile tools in a shop. In addition to making all the standard saw cuts, it can operate as a drill, a router, a saber saw, a sander, or a grinder with the right attachments. The instructions that come with the saw should explain how to use such accessories.

Although it costs about the same as a table saw, a chief attraction of the radial arm saw is that it fits snugly along one wall of the shop rather than in the middle, like a table saw.

The radial arm saw consists of a base, a worktable, a support column for the arm, and the arm itself, which contains a track on which the motor and saw slide back and forth. When out of use, the saw is moved to the rear of the arm, near the support column. Boards to be cut are placed on the table, and the saw is pulled out along the arm to make the cut.

The motor and attached saw can be rotated 360 degrees for miter or rip cuts, or can be tilted on its side for bevel cuts. A gauge calibrated in degrees located next to the motor is used to align the saw blade. For crosscut operation the saw blade rotates with the teeth spinning downward, which presses the board being cut against the fence at the rear of the table. The fence is at right angles to the saw for straight crosscuts.

Unlike the metal top of a table saw, the table for a radial arm is made of soft particleboard. In normal operation the saw blade will lightly cut this surface in numerous places as it slices through the wood. Some operators cover the tabletop with ¼-inch hardboard held in place with recessed flathead screws. When this becomes severely scarred, it can be replaced more easily than the entire table.

The table must always be square with the saw blade. Different models have different means of adjusting the table, so check the instructions carefully. A final check can be made with a steel framing square resting on the table and against the saw blade. But make sure the square rests against the gullets in the blade and not against the teeth, which may be offset.

## Buying a Radial Arm Saw

One of the first questions that comes to mind when you are looking for a big shop saw is whether to buy a table saw or a radial arm saw. The radial arm saw is popular in home shops because it fits against one wall, whereas a table saw must be placed in the middle of the shop so the wood can be pushed over it. With a radial arm saw, the stock is placed in front of the saw and the saw is moved across it.

One limitation of the radial arm saw is that it won't cut large sheets of plywood, since there is only about two feet between the arm support and the fully extended blade. That limitation is easily overcome by cutting the plywood with a circular saw instead.

As with other power saws, a major consideration in buying a radial arm saw is how big and how powerful it should be. Most 8-inch radial arm saws are designed for cabinetmakers. For general shop use, select the 10-inch saw. If you know you will be using your saw a great deal, you might want to go to the 12-inch saw. Saws larger than 12 inches are for commercial use only.

It is difficult to recommend a specific horsepower. Increased horsepower means increased cost, and you must decide for yourself what you can afford. Remember that the smaller the horsepower, the harder the motor has to work and the sooner it will wear out. It is in your best interests to discuss with the salesperson the type of work you will be using the saw for. You should also find out what the actual horsepower is, as opposed to the developed horsepower. The motor may develop much more horsepower in its initial surge than it will under a load.

Another thing to check is the sturdiness of the stand. It should be well braced and should be made of a heavy-gauge metal, since the added weight helps stabilize it.

Be sure that the saw has a brake on the blade. A free-wheeling blade is very dangerous. Some saws have an automatic brake that stops the blade when power is cut; others have a manual brake. Either one will do the job.

Finally, find out exactly what is included in the price. Some saws are sold without such useful items as the motor.

### Radial Arm Saw Blades

Blades for the radial arm saw are the same as those with table saws, which are discussed on page 34. However, the size of a blade on a radial arm saw does not always correlate with how deep a cut it will make. This depth is affected by the design of the radial arm. Some 10-inch radial arm saws, which will take a blade up to 10 inches in diameter, will cut only 3 inches with the blade set vertically. Others will cut up to 3⅝ inches. However, all 10-inch saws will cut a 2 by 4 when the blade is set at a 45 degree angle.

## Basic Cutting Techniques

All of the basic cuts on the radial arm saw are made by rotating the saw blade for the miter cut, or tilting it for the bevel cut, or a combination of both.

**Crosscut.** The radial arm saw is ideal for making crosscuts rapidly and accurately. Simply place the board to be cut against the fence and align the cutting mark with the blade. Remember to cut on the waste side of the line. Before starting the saw, pull the blade forward to the cutting line to check your alignment. Raise or lower the blade so it just cuts through the stock.

When crosscutting numerous pieces the same length, tack a piece of scrap to the table at the right distance from the blade to serve as a stop.

**Rip cut.** When making a rip cut, it is very important that you turn the blade so that its rotation is up and against the stock being pushed through. If the blade is rotating down, as it does in crosscutting operations to hold the board against the fence, it may suck the board through and possibly pull your hand with it.

When the saw is turned in the proper direction, depending on whether you are feeding from the right or the left, press the back edge of the board against the fence and lock the saw in position over the cutting line. Raise or lower the blade so it just cuts through the stock, then make the rip.

**Miter cut.** The miter cut is easily made either by turning the blade to a predetermined angle, which is found on the gauge, or by turning it to align with the angle marked on the face of the board to be cut.

**Bevel cut.** As with the miter cut, the saw can be tilted to a predetermined angle or tilted to match the angle marked on the edge of the stock to be cut. Once the stock is held against the fence and aligned under the blade, pull the saw across it to make the cut.

**Bevel miter cut.** Set up the blade for the proper bevel and miter. Make sure the stock is firmly against the fence, and not angled out slightly by a piece of scrap wood or sawdust.

**Bevel rip cut.** Pull the saw out on the arm and rotate it 90 degrees. Remember that, as in a rip, the blade must rotate up and against the stock, not down as it does for crosscutting. Tilt the blade to the desired bevel, lower the saw so it will cut through the stock, and make the cut.

**Dadoes, rabbets, grooves, and tenons.** As with the table saw, these are best made with a dado blade. They are executed in the manner described on page 36, except for the fact that the proper length is found by raising or lowering the blade above the stock, rather than below it.

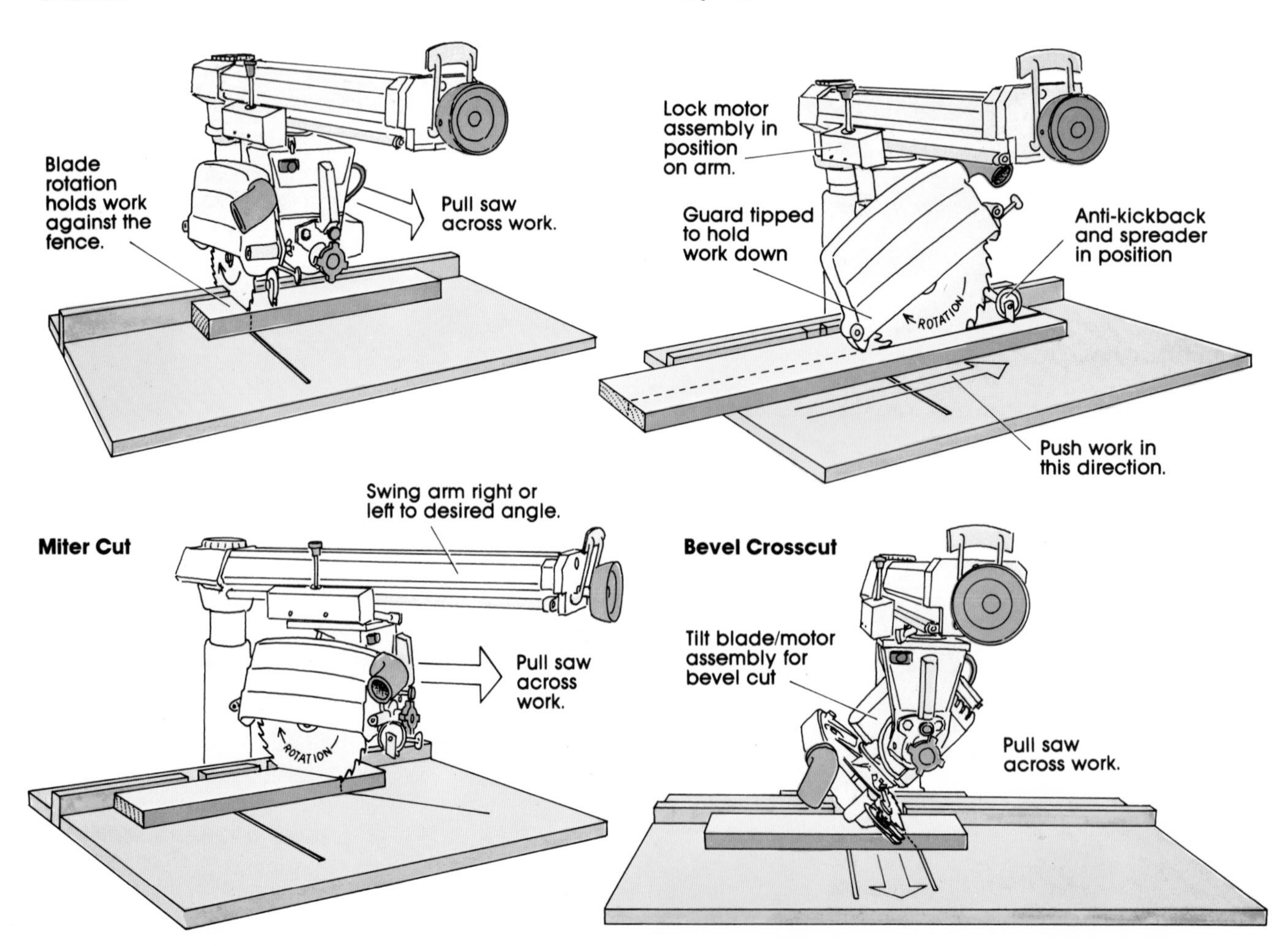

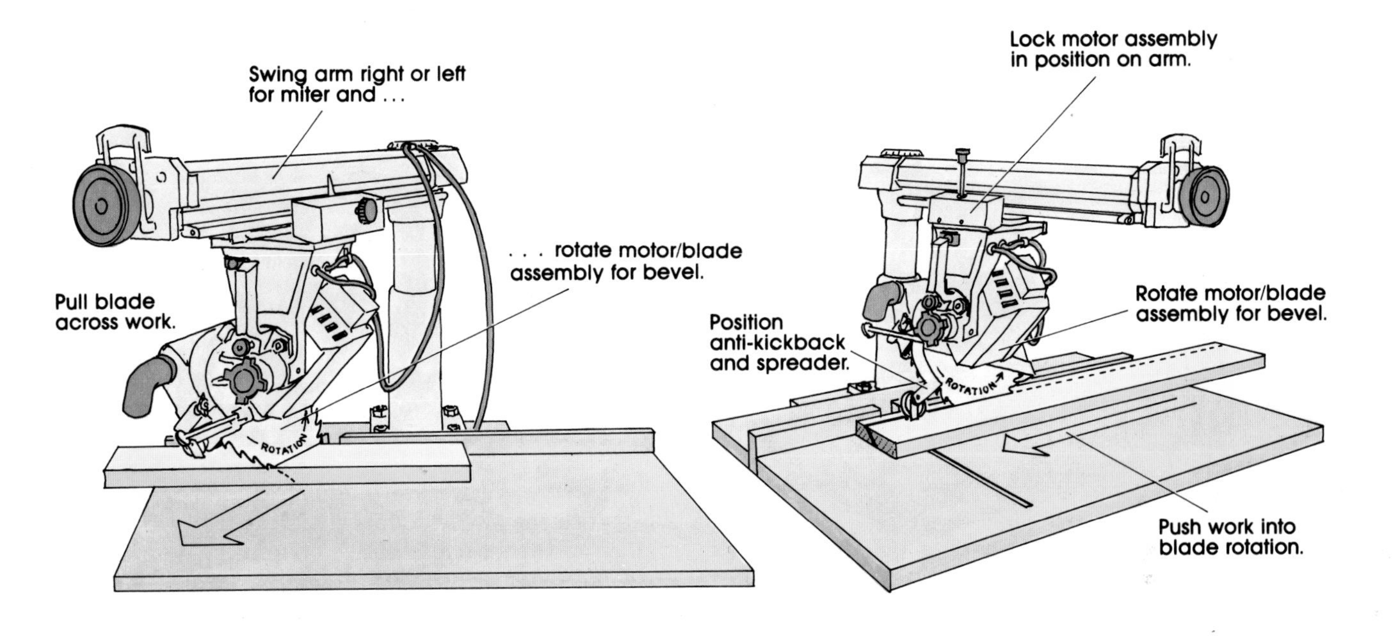


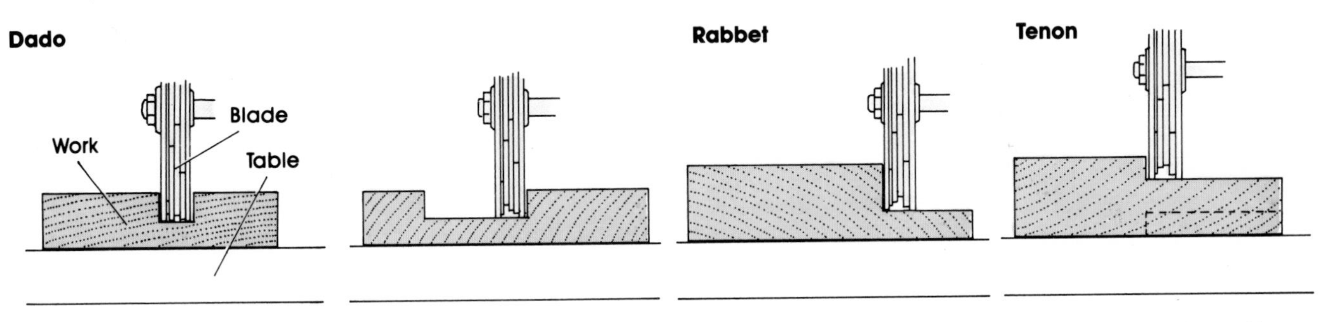


For a dado or groove from 3/16-inch to 13/16-inch, set the blade to the width you want, set the saw to the desired depth, and make one pass.

For a dado or groove wider than the blade, make several passes.

A rabbet or lap joint cut can be cut by making one or more passes with the blade set to the desired depth.

A tenon is cut by making several passes, then turning the work and making an equal number of passes on the other side.

## Tips on General Operation

■ When you change a radial arm saw blade, always pull the power plug first. Note that the nut that holds the blade has a left-hand thread and is loosened by turning it clockwise.

■ Watch out for "heel cutting." As you cut a board, see if the blade is kicking up more splinters on one side than the other. If it is, the blade isn't traveling parallel to the radial arm. There are normally three set screws on the back of the motor housing that are turned to make any adjustments. Check the instructions on your particular saw.

If you have to cut lumber thicker than your saw will handle in one pass, mark the wood on both sides, then make the first cut. Turn the wood over and make the second cut, being careful to align the second cut properly over the first.

## Radial Arm Saw Safety Rules

■ In crosscutting operations, make sure that the blade teeth are moving down into the wood to hold it firmly against the fence.

■ In ripping operations, make sure the blade teeth are moving up and into the wood, pushing it back toward you rather than sucking it through the saw.

■ Don't stand directly in front of the saw while working.

■ Always push the saw to the rear, out of harm's way, when you finish making a cut.

■ Make sure electric cord is completely out of the way.

■ Check that the power is grounded.

■ Keep the floor around the saw free of debris.

■ Keep all locking/clamping devices on the saw tight.

■ Don't use damaged blades.

■ Always wear safety goggles.

# DRILLS

Parts of a Hand Drill

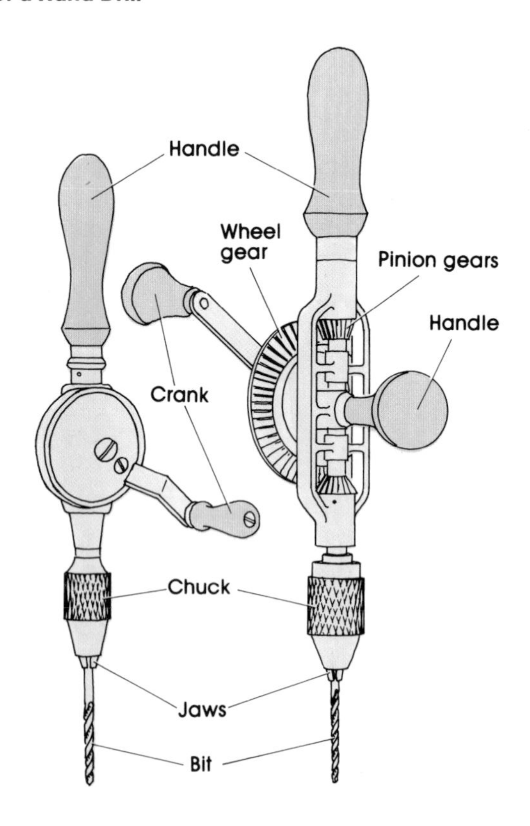


## Hand Drill

Even in this seemingly all-electric world, there is a place for hand drills in any shop. Anyone who loves working with wood will have several different types for different jobs.

The hand drill is generally used for making small, precise holes for screws, bolts, or dowels. Its bit fits into the clamping jaws inside the chuck and is tightened by turning the crank until the jaws are snug. The bit must be straight, tight, and held evenly by the jaws. To tighten the jaws, hold the chuck firmly to keep it from moving and turn the crank hard. To remove the bit, hold the chuck and turn the crank backwards.

When drilling in hardwood or metal, first make a small starting hole with a nail set, center punch, or awl (see pages 60–61). This will prevent the bit from slipping about on the surface. Hold the drill either vertically or horizontally when drilling. A hole drilled at a slant will result in the screw not setting properly.

With hand drills you cannot reverse the cranking action to back the bit out of the hole. This will only open the jaws. To remove the bit from the hole, keep cranking forward while you pull back on the drill. Never rock the bit back and forth in the hole, or you will bend or break it.

When you must drill to a precise depth, measure the depth on the bit and mark it with a piece of tape.

## Push Drill

The push drill is used for small jobs around the house or light cabinetwork. It enables you to drill a hole with a regular bit and then change to a screwdriver bit to drive the screw in quickly. Most models also have a reverse action for removing screws.

## Brace and Bit

The brace and bit is used for jobs that require larger bits and more power than a hand drill or push drill can provide. Because more turning pressure is applied with a brace, it has a different type of jaw and takes a different type of bit.

A brace has a V-groove in the jaws and requires an auger bit with a square tang. When placing the bit in the jaws, make sure the edges of the tang slip into the grooves before tightening the chuck. The chuck is tightened by holding it firmly while turning the brace handle.

The first time you use a brace and bit, you may find that it wobbles back and forth. Until you gain experience, place a try square beside the bit and use that as a guide to keep the drill vertical. When you bore large holes in narrow pieces of wood, too much pressure may split the board. To avoid this, work slowly and with only moderate pressure.

Auger bits have a screw feed on the tip that helps pull the bit into the wood. When planning to countersink screws, always drill the larger countersink hole first. If you drill the smaller screw hole first, there won't be any wood for the screw feed to grab.

**How to Avoid Splintering**

When you drill through a board with a hand drill—and to a lesser extent with a power drill—the back side of the board will splinter as the bit emerges. One way to prevent this is to clamp the board to a piece of scrap. The bit then moves cleanly into the piece of scrap without tearing out the back of the good board.

Another method is to stop as soon as you see the tip of the feed screw emerge at the back. Using this small hole as a guide, finish the hole by drilling from the back.

Parts of Hand Drills

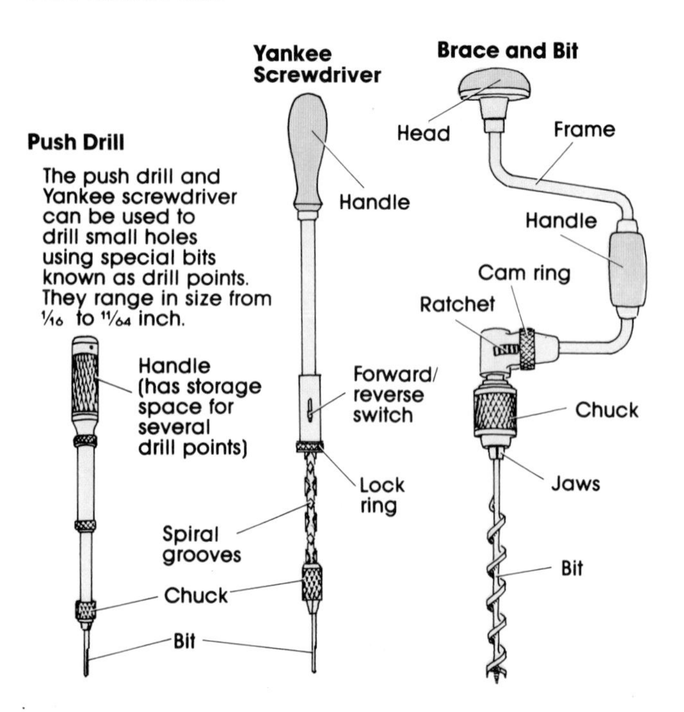

**Parts of a Power Drill**

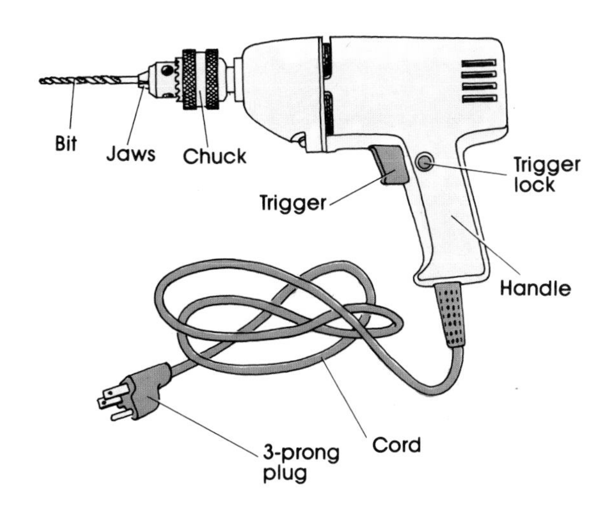


## Power Drill

If you can afford only one power tool, the electric drill has to be the first choice. With the right accessories, this simple drill will operate as a grinder, saber saw, circular saw, drill press, power plane, sander, chisel, screwdriver, wrench, or paint mixer.

Power drills are designated according to the size of the largest drill bit shank the chuck can hold. These sizes are commonly ¼ inch, ⅜ inch, or ½ inch. Larger and more powerful drills are for commercial use only.

The smaller the drill, the faster its revolutions per minute. A ¼-inch drill, for instance, might operate at 2200 rpm with no load, whereas a ½-inch drill of the same make might operate at 700 rpm. High rpm is acceptable for small jobs but will strain the motor and tend to burn it out on heavier jobs.

Before you purchase your first drill, consider how much carpentry work you will be doing. If it is just small cabinetwork, you can get by with a ¼-inch drill, but a ⅜ inch is much more versatile. If you plan on building a cabin or large deck, you should buy a ½-inch drill.

The ⅜-inch drill with reverse gears and variable speed is generally a good choice. The variable speed allows you to tackle heavy jobs without overworking the motor and also makes it possible to drive screws at slow speeds. The reverse switch makes it easy to remove screws.

Drill prices vary considerably, but you have to pay for quality. As a rule of thumb, the longer the drill cord, the better the drill. A short cord is a nuisance because even the simplest job means locating an extension cord first.

## Using a Power Drill

Using a power drill is quite simple. However, when you use one for the first time, concentrate on holding the drill steady and perpendicular. If the drill wobbles back and forth, the bit can snap, and if the hole is not at a right angle, the screw head will not fit flush.

After you select the right bit for the job (see page 42), use a nail set or awl to make a small starting hole in the wood. The hole prevents the drill bit from skipping across the wood when you start. Next, place the bit in the hole, holding it with light pressure, and start the drill. Experience will teach you just how much force to use, but push hard enough to keep the bit moving without severely lugging the motor. On larger projects, it's a good idea to withdraw the bit periodically to remove waste wood from the hole. Removing the bit from the hole is easy if you keep the drill running. It is not necessary to use the reverse gear when you remove the bit unless it is stuck.

If you must drill at an angle in an awkward position, you can drill a more accurate hole by first drilling through scrap wood at the desired angle and then clamping the scrap board onto the spot where you want the hole.

When you drill very hard wood, it's a good idea to drill a small pilot hole before drilling the larger one. With less wood to remove there is less resistance to the bit and less chance of the wood splitting.

### Countersinking Screws

To *countersink* a screw means to set the head below the wood's surface by drilling a shallow hole to accommodate it. To determine how big this hole should be, drill a test hole in a piece of scrap, then turn the screw upside down and see if the head fits. Sink the screws about ⅛ inch below the surface, then cover them with wood putty.

Combination bits, which drill the pilot hole, the shank hole, and the countersink in one operation, are now widely available in hardware stores. They come in a variety of sizes to match different screws.

### The Power Drill As a Screwdriver

If you are making a cabinet or table or anything that may require numerous screws, it's a real time saver to drill all the screw holes first and then use the power drill to drive them. To do this, select a screwdriver bit that matches and fits the screw slot. All sizes and types are available. The best type of drill to use is a variable-speed drill, which enables you to start the screw at a very slow speed then increase the speed to drive it home. A second choice is a 3-speed drill operated at the slowest speed. An ordinary drill with one constant speed cannot be used because it turns too rapidly.

When you drive screws in hard or green wood, you can make your job easier by first twirling the threads on a bar of soap or wax.

### Drilling into Concrete

Construction work sometimes requires drilling into concrete. A common example is drilling into the side of a concrete house foundation to fasten a board, or "stringer," as part of a deck. To do this you need at least a ½-inch drill. The motor is powerful enough to keep working without overheating, and it turns at a slower speed than small drills. Higher speeds will quickly burn up the bits. You will also need carbide-tipped concrete-boring bits especially made for this chore.

Boards are normally fastened to concrete with expansion bolts, so select the size you need first and then choose the drill bit to match.

If you find the going too tough for your drill, try working

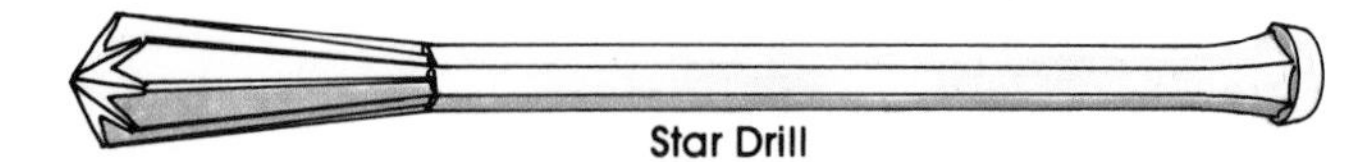
Star Drill

awhile with a star drill, which is a hardened cold chisel with a fluted end. Use a five-pound hammer to hit it, and turn the star drill slightly after each blow. Alternate it with the masonry drill. You should always wear goggles when using a star drill.

## Bits and Drill Accessories

There are two basic types of drill bits: auger bits, which are used with hand drills, and twist bits, which are used with both hand and power drills.

### Auger Bits

There are two types of augers, the *solid shank* auger and the *twist* auger. They look fairly similar, but the twist auger has more and deeper spirals than the solid shank auger, which is stronger and preferred for heavy drilling.

Both types of auger bits have a screw feed on the end, which looks just like a screw. This screw feed literally pulls the bit through the wood while the cutting edges just behind it remove the wood. The threads on the screw feed differ according to the type of drilling you must do. For hardwood, use a fine thread screw so the bit will not work its way too rapidly through the wood and possibly split it. For softwood, choose either a medium or coarse thread screw.

The cutting end of the bit, right behind the screw feed, has two spurs that literally slice the wood in thin shavings as the bit rotates. The other end of the bit, which fits in the chuck of the brace, has a square, tapered tang. This is locked in a pair of V -grooves in the chuck so that great torque, or turning pressure, can be applied without the bit slipping.

Auger bits range in diameter from ¼ inch to 1 inch. They increase in $\frac{1}{16}$-inch gradations, and the sizes are stamped on the drill in sixteenths of an inch. For example, a ½-inch bit would be stamped "$\frac{8}{16}$."

**Auger Bits for a Brace**

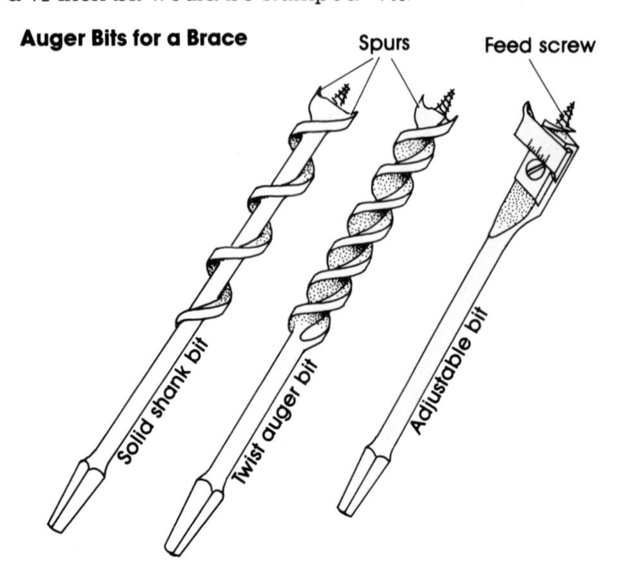


**Drill Bits**

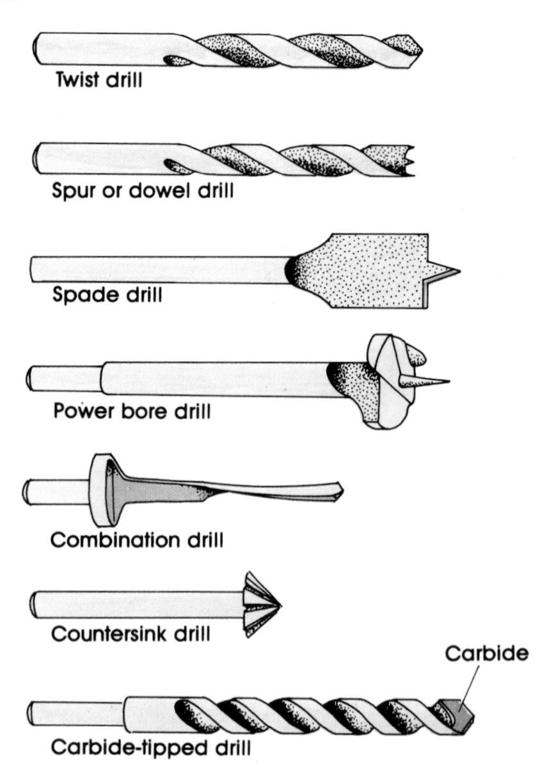


### Other Power and Hand Drill Bits

The most widely used bit for power or hand drills is the *twist bit*. It can be used for drilling either wood or metal, and normally ranges in size from $\frac{1}{16}$ inch to ½ inch. The sizes are stamped on the side in $\frac{1}{16}$-inch gradations. The quality of steel that is used in different brands of bits is not always indicated, but you can usually judge by the price. Buy the best you can afford so you will not have to be constantly sharpening or replacing the cheaper bits.

Some of the other most widely used bits are illustrated above.

The *spur bit* is a miniature version of the auger bit, with tiny spurs that cut quickly through wood. It should only be used on wood. Sizes range from $\frac{1}{16}$ to 1 inch.

The *spade bit* works very well for drilling wood. It gives a fast, clean cut and the tips can be readily sharpened. Because this bit works best at high speeds, it works much better in a power drill than in a hand drill. Sizes range from ½ to 2 inches.

The *power bore* works in a manner similar to the spade bit, but gives a slightly finer cut. Sizes range from ¼ to 1 inch.

The *combination bit* drills the small pilot hole for the screw tip, the wider hole for the shank of the screw, and the countersink for the screw head in one action. It is excellent for cabinetwork and is available in a variety of sizes to match different types of screws.

The *countersink bit* has a beveled face that allows you to enlarge the top of a screw hole so the head of the screw can be sunk flush with the surface of the wood.

The *carbide-tipped bit* is tipped with extra hard carbide steel and is used to drill concrete or stucco. Sizes range from ¼ to 1 inch.

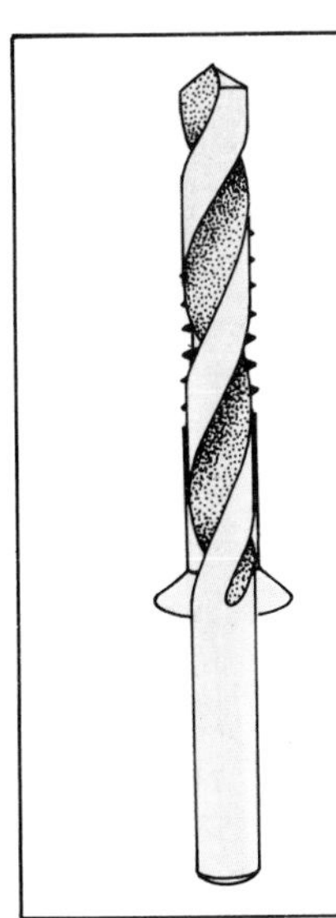

**Matching Bit to Screw**
Although there are charts that tell you what size screw goes with what bit, woodworkers commonly use the following method for making a selection: place the screw you want to use directly behind a bit that appears about right. You should just be able to see the threads of the screw on either side of the bit. If you can't see the threads, the bit is too big; if you see part of the main shaft of the screw, the bit is too small.

## Power Drill Accessories

The wide variety of accessories for a power drill make it a nearly complete workshop by itself. The number of attachments you buy depends on your needs, but you should be aware of what's on the market. If you plan to use numerous attachments, see how many are available for a particular model before buying the drill.

One of the first attachments invented for a power drill was the *sanding disc*. It can be either flexible or rigid. The flexible model is commonly used for freehand sanding or buffing, while the rigid disc is used with the drill attached to the workbench. For buffing and polishing, a variety of covers are available that simply slip over the same flexible rubber disc that holds the sandpaper.

The *drum sander* is used for working on curved edges and in awkward places. Be sure to check out the Sureform drum sander. It does the same job faster than other drum sanders, though more roughly, and can be used for shaping.

The *wire brush* attachment serves many purposes, including cleaning old and rusted metal tools or removing chipped and flaking paint from wood. If you want to strip furniture without hours of hand work, try using the *stripper* attachment.

Any shop can find uses for a *grinder* attachment. It comes in several levels of abrasion and can be used for sharpening, shaping, or cleaning. The simplest way to use a drill as a bench grinder is to attach it to a drill stand. A drill press attachment is worth buying if you have a lot of precision drilling to do. Because it is light and portable, you can use it in places a standard drill press would not reach.

The *circular saw* attachment allows you to cut up to 2-inch stock, but because the blades are normally limited to 6-inch diameters, you cannot cut a 2-inch board at a 45 degree angle.

The *saber saw* attachment locks securely to the drill and allows you to work with one hand while steadying the wood with your other hand.

The *electric plane* attachment is worth looking into if you have a lot of planing to do. It will cut a width of up to 2 inches and a depth of up to ⅛ inch on each pass.

The *hole saw* is a specialized device, but it is extremely useful when you need it. It cuts out a disc rather than boring a hole, and is most commonly used for cutting holes in doors to install doorknobs. It comes in a variety of sizes.

The *bit extension* is for drilling holes in tight places that a drill won't reach. Be careful not to apply too much pressure; if the extension is bowed while you are drilling, a severe whipping action can result.

*Right-angle drill attachments* are used for working in confined areas. Electricians commonly use them to drill holes through studs to run the house wiring.

**Power Drill Accessories**

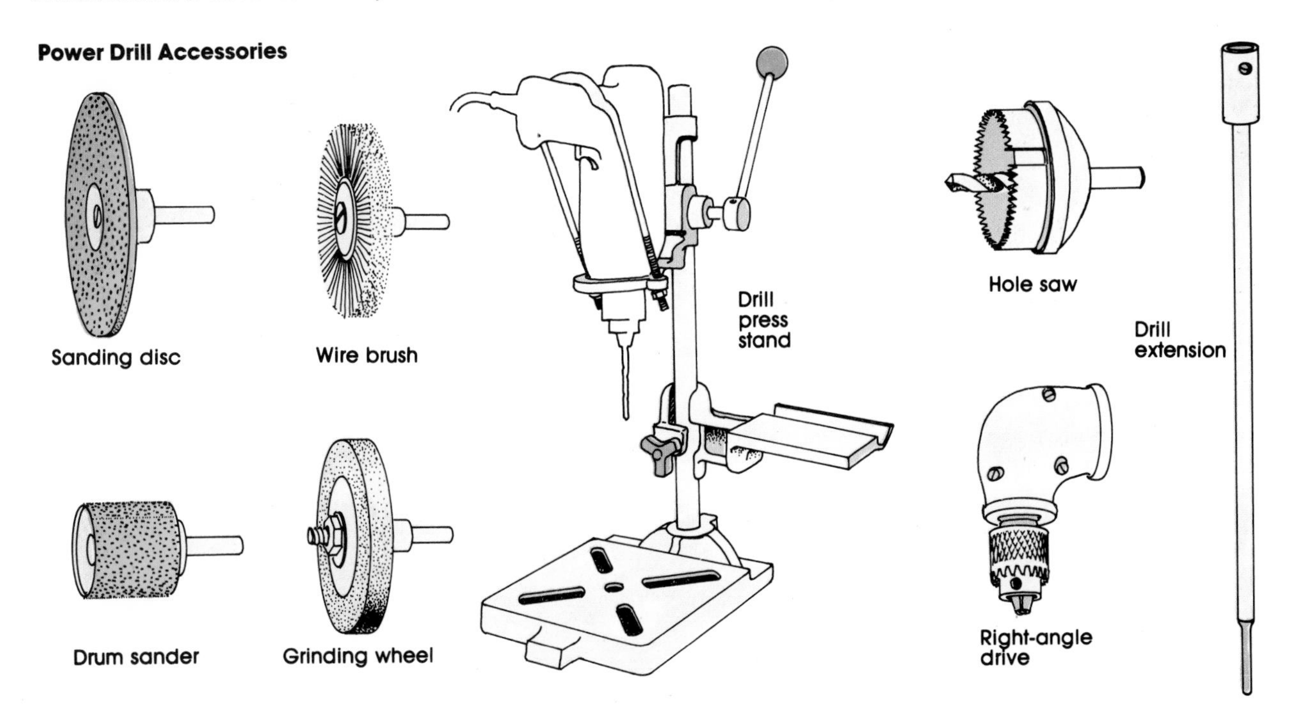

**Parts of a Router**

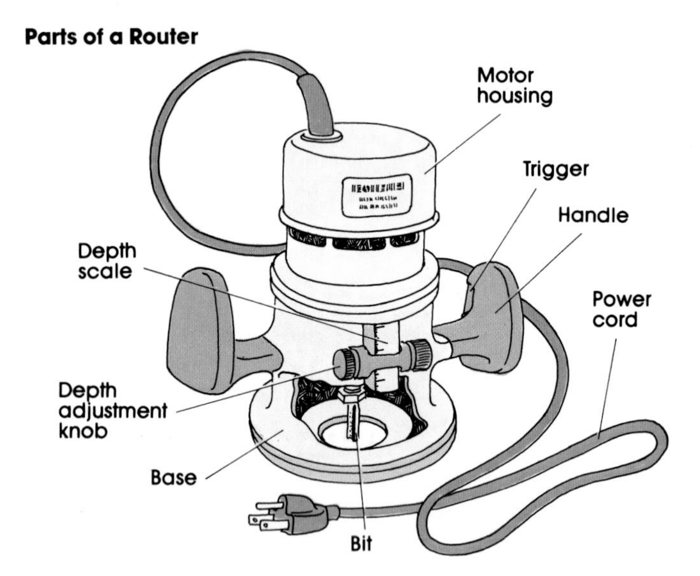


Once you use a router you'll wonder how you ever got along without one. With the wide choice of bits available, you can make fancy edges on tables or cabinets, cut dovetails in minutes, professionally trim plastic laminates on countertops, make moldings, and perform a host of other tasks.

The router consists of a motor that drives a shaping bit at tremendous speeds—up to 27,000 rpm, compared to 2,200 for an average drill. Its motor can be raised or lowered to adjust the cutting depth of the bit. A scale on the housing is used to set the depth, which on some models can be as fine as 4/1000 of an inch.

Routers range in horsepower from ½ hp to 3 hp. The smaller the motor, the harder it has to work and the faster it will wear out. Since you will probably use the router more than you think, choose at least a 1 hp model.

## Using the Router

It's important to get the feel of working with a router. As you make a cut, you will hear the motor slow down somewhat when the bit contacts the wood. It's important not to cut too rapidly and lug the motor, which can burn it up. At the same time, don't go so slowly that you overheat the bit and remove its temper.

The router excels at making rabbets, dadoes, and grooves. Be sure to select proper bit (see page 45). Then, adjust it to the correct depth by placing the router base on the board to be cut and lowering the motor until the bit is aligned on the edge of the board at the desired depth. The depth of the cut should not exceed two-thirds of the board's thickness. During the cut make sure the board is firmly positioned or clamped so it will not move.

For greater precision in making cuts, it's best to use a guide. You can either purchase an adjustable one or clamp a straightedge next to the router base.

A handy guide you can make yourself consists of a simple wooden U. Screw and glue two pieces of 1 by 4 together in the shape of a U and make sure they are at right angles. Use a square to check. When using this guide, set the top of the U over the edge of the board and clamp the leg of the U to the board next to the router.

If you need to rout an area wider than the bit, make several passes, sliding the guide down the board for each new cut.

For decorative work of your own design, you can make your own template, or pattern guide, from a piece of ¼-inch plywood. When you use a template, a *guide-bushing* must be screwed to the base of the router. First, draw your design on paper and then transfer it to the plywood by using carbon paper. Allow for the off-set between the router bit and the edge of the guide-bushing. Place some scrap wood under the piece to be cut and adjust the router so the bit is 1/16 inch below the plywood. Then cut it freehand with the router. Any irregularities can be repaired by sanding out the rough spots or filling gouges with plastic wood. Next, with the template clamped in place, lower the bit to the desired depth and follow the template to cut the actual workpiece. The template is, of course, reusable and allows you to repeat the pattern precisely.

## Router Bits and Accessories

Bits are designed to make either grooves or shaped edges. There are many different kinds and they can be expensive, so it might be wise to buy a kit with a basic selection and then buy the more specialized bits as you need them. Bits are made either of high-speed steel or much harder tungsten carbide. If you are doing a lot of work on plastic laminates, the carbide bits are a good choice for long wear.

Bits can be sharpened with the aid of a sharpening attachment. It results in precise work, but different sharpening grinders must be bought for the different bits.

While the router itself is worth the money you pay for it, the accessories can quickly run up the cost. But if you can make a lot of use of a particular accessory, you will quickly recover the cost in time saved.

If you are doing a lot of cabinetwork or decorative work, the shaper table may be a worthwhile investment. The router is fixed underneath the table, and the pieces of wood are guided along the top. The table should be sturdy. Many models have grooved tops to prevent sawdust from building up under your workpiece. While router

**Homemade T-guide**

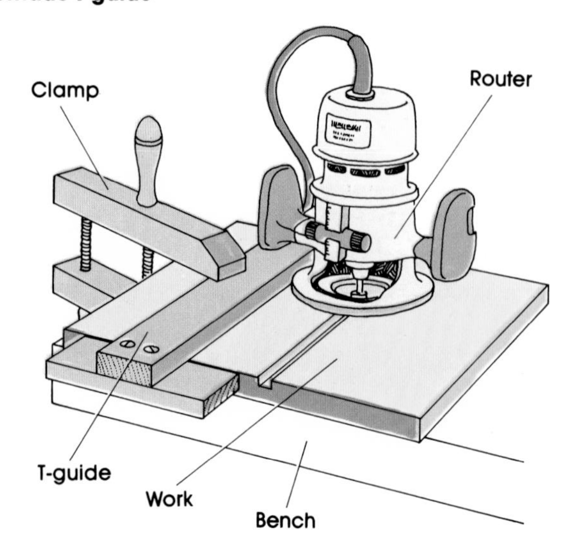

Router Bits

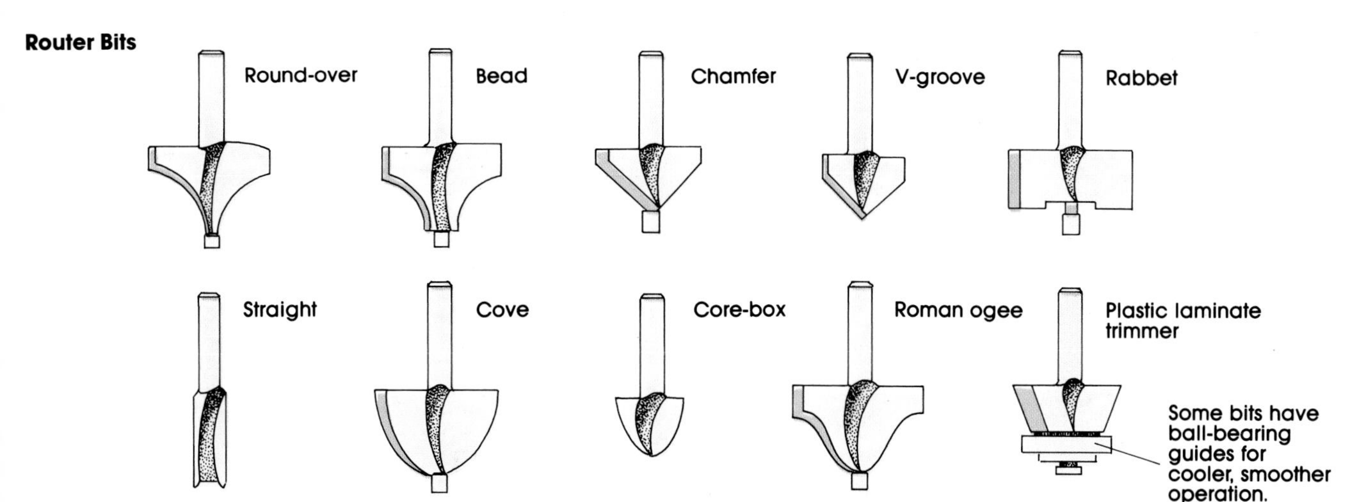


bits can be used on shaper tables, it is worth buying shaper cutters for extensive work.

If you make a lot of drawers or fine cabinet pieces, a dovetail template is an ideal accessory. This template, coupled with the dovetail bit, allows you to do in minutes what would otherwise take hours of laborious hand work.

Electric planes are available for several different models. They will generally make a cut 2 inches wide and ⅛ inch deep and are very useful if you have a lot of doors to work on.

Other widely used router accessories include templates for routing out hinge butts on doors and jambs, for making mortises for door locks, and for routing out space for treads and risers in stair stringers. Another specialized attachment is designed for trimming veneer and is especially useful for working on formica countertops.

## Router Safety Rules

- Make sure the router is turned off before plugging it in.
- Always unplug the router while changing bits.
- Always wear safety goggles.
- Hold the router firmly and with both hands while working with it.
- Make sure the piece you are working on will not move.

Shaper Table

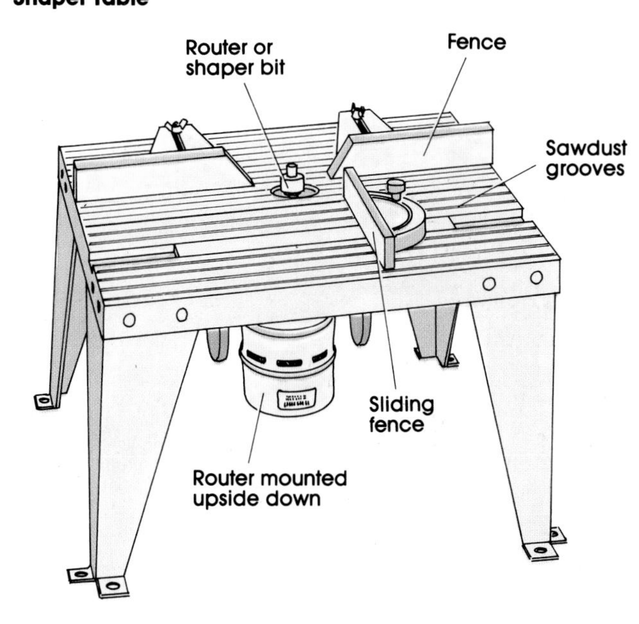


Dovetail Template

Dovetail Joint

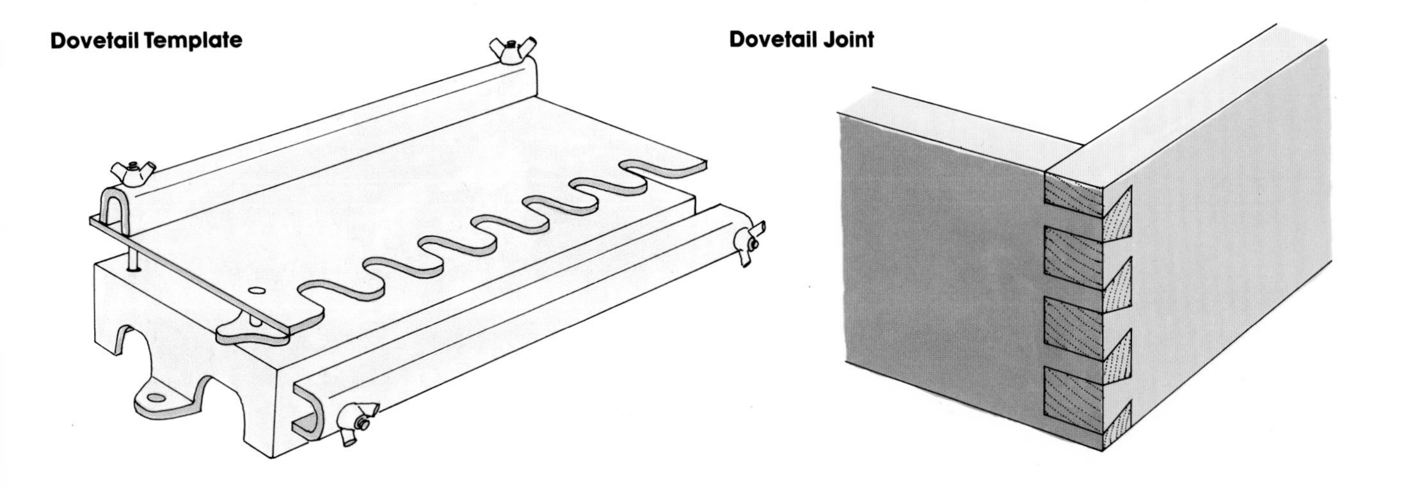

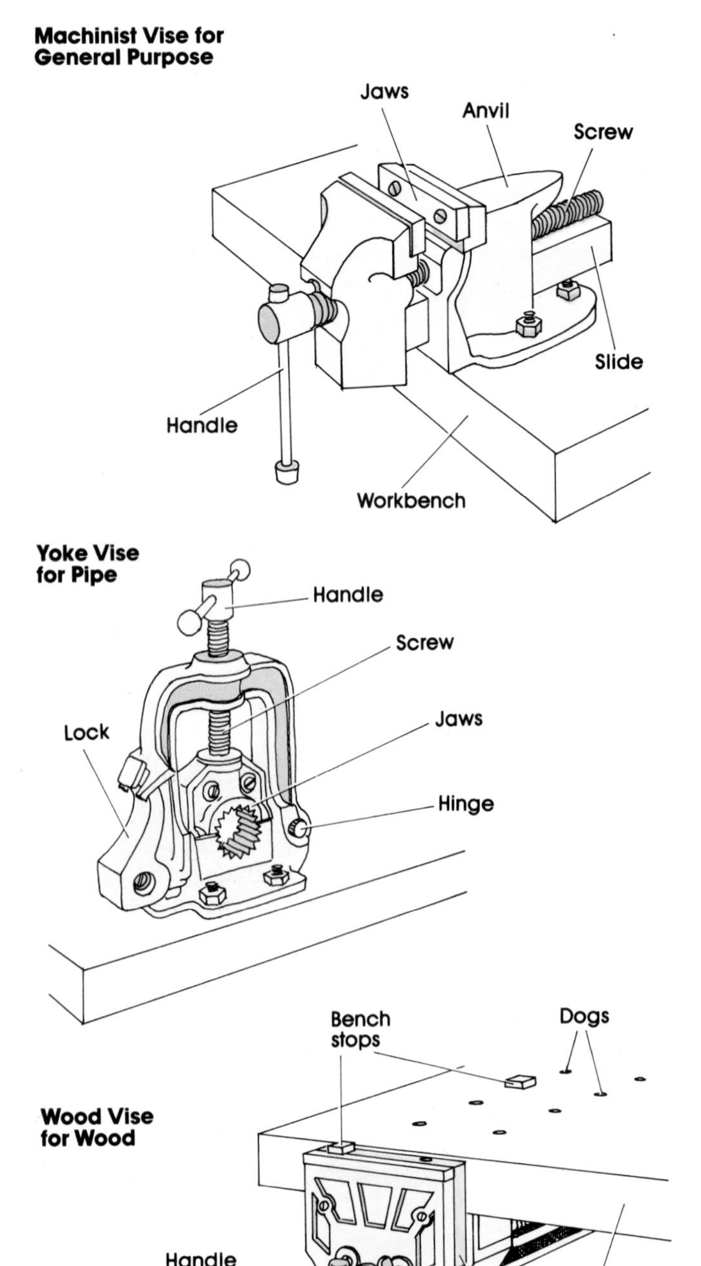


A vise is used to hold whatever you are working on securely and thus frees both your hands. It should be securely bolted to the workbench, usually at one end where it's out of the way, and should extend far enough out so the edge of the bench will not interfere with the workpiece.

On many models, the clamping jaws are scored to give a firm grip, but this can also damage wood. Either clamp the wood between two pieces of scrap, or, if there are screw holes in the jaws, screw pieces of wood to the jaws. Sheet metal, bent at right angles, can also be slipped over the jaws to provide protection.

The basic styles of vises for home shops are illustrated on this page.

The *machinist's vise* is the standard shop vise. It comes in several different sizes and is used for holding everything from wood to pipe. The better models can be swiveled and have a flat anvil space on the back side for light metalworking.

The *yoke vise* is primarily used for holding pipe without crushing it, as a machinist's vise might do. It comes in many different sizes. Unless you intend to do a lot of pipe work, choose a smaller style that can be clamped to the workbench or quickly set up on a sawhorse at a building site.

The *wood vise* is designed to fit flush with the top of the workbench, which is often very convenient. It has broad jaws with predrilled screw holes for mounting scrap wood so the workpiece will not be marred. Use flathead screws that are countersunk in the wood. Many wood vises have an adjustable dog on the top, which, in conjunction with a bench stop or the workbench, is used to hold large pieces of wood in place.

## Clamps

When two or more pieces of wood are glued together, clamps are generally used to hold the pieces firmly until the glue dries. (See page 48 for a discussion of gluing techniques.) Human beings have invented hundreds of ingenious ways of putting wood together, and there are many types of clamps available. Some of the basic types are illustrated on this page.

The *C clamp,* named for its shape, is a standard item in any workshop. As with many other clamps, you should use scrap wood under the jaws to prevent damage to the wood.

The *bar clamp* has one fixed end, while the other end slides back and forth on a miniature I-beam and then screws down tight. It provides more range than the C clamp.

The range of a *pipe clamp* is limited only by the length of the pipe. One end screws to the threaded end of ½- or ¾-inch pipe, while the other end slides back and forth. You supply the pipe. Lengths of five or six feet are adequate for most jobs. This clamp is useful in many situations where several boards must be clamped together, such as for a tabletop.

The *hand screw clamp* is specifically designed for woodworking. The wooden ones are made from 2-by-2 hardwood and will not damage the workpiece when tightened. They are adjustable for holding angled pieces, but be sure they are tightened with equal pressure on both ends.

The *spring clamp* is like a big clothespin and is suitable only for light work.

The *web clamp* is widely used for wrapping around the legs of a chair to cinch them tight while glue dries.

The *miter clamp* can hold two pieces of wood together at right angles. It is commonly used for making picture frames.

A *three-way clamp* is specially designed to hold two pieces of wood tightly in place from three angles. This is necessary, for example, with countertop laminates.

**Clamps**

C clamp

Bar clamp

Pipe clamp

Hand screw clamp

Three-way clamp

Spring clamp

Web clamp

Wrench

Web

Ratchet nut

Release lever

Miter clamp

## Adhesives

When you join two pieces of wood together, the bond will always be stronger if you use glue in addition to the nails or screws. The only projects you wouldn't want to use glue with are those you plan to dismantle later, such as bookshelves or tables that must be taken apart at moving time.

Glues are used almost exclusively in finish carpentry. (See page 21 for a discussion of wood joints.) A notable exception is the application of adhesive to the tops of floor joists to which plywood subflooring is to be nailed. This helps prevent squeaking.

Six of the most commonly used glues, what to use them on, and their advantages and disadvantages are covered in the chart on the next page.

Before you start to glue two pieces of wood together, consider the following questions:

■ What kind of wood are you using? The harder the wood, the more problems you may encounter in attaining a good bond. The softer woods absorb glue more readily along the surface of the wood and thus hold better. In such cases, the glue is often stronger than the wood—the wood will break before the glue will.

■ How much pressure is needed for a good bond? Pressure from clamps is necessary for all wood gluing. It forces the glue to spread uniformly, pushes out air bubbles, and presses the two wood surfaces tightly against each other. To determine how much pressure is needed, use this rule of thumb: when you are working with a thin glue, such as white glues, use only moderate pressure; with thicker glues, such as plastic resins, use heavy pressure.

■ What types of clamps are required, and how many? This depends largely on your project. The main thing to remember when you clamp wood is to maintain even pressure over the entire area. This generally requires two or more clamps. Always put a piece of scrap wood on both ends of the clamp so you will not mar your workpiece. When you glue long pieces together, such as for a tabletop, use at least three bar or pipe clamps. The two clamps near the ends should be across the top, while the clamp in the

**Using Pipe Clamps**

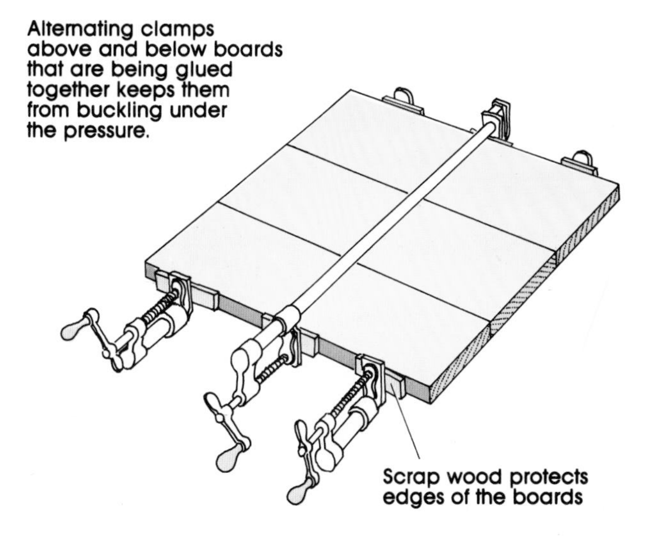

middle should be underneath. This helps prevent buckling.

■ How long does the glue take to set? This depends on the type of glue you are using and the air temperature of your work area. Check both the chart and the glue directions for drying times. Removing the clamps too soon can ruin your work.

### Gluing Techniques

Before you apply any glue, wipe the surface spotless with a soft, dry rag. Any wood particles that remain will only act as tiny wedges to hold the pieces apart.

Do not rough up the wood prior to gluing. The surface will be dry, but underneath it may be damp and that moisture will hinder a tight bond.

At all costs try not to glue two pieces of green (freshly cut) wood together. Not only will the moisture content interfere with the bonding process, but also as the wood dries it will shrink away from the bond.

### Glues

Selecting the right glue for the job can be confusing. The following chart provides basic information on the advantages and disadvantages of six major varieties of glue.

|  |  |
|---|---|
| **Flake Animal** | Good for light colored joints and filling cracks and gaps in joints when working with wood and furniture. Must be kept hot so it is not suited for quick jobs. |
| **Liquid Animal & Fish Glue** | Easy to use, strong and light colored. Resists heat and mold but not water. Use for general wood gluing but warm first if using in a cold climate. |
| **Plastic Resin** | Light colored, strong and water resistant. Good for work that may get damp. Must be mixed and only works well on projects that are clamped overnight. |
| **Powdered Casein** | Can be used in cold places. Good for general woodworking and oily woods such as teak, yew, and lemon. Strong and somewhat water resistant. Must be mixed fifteen minutes before using and cannot be stored indefinitely. Stains dark woods and is subject to mold. |
| **Resorcinol** | Best used in warm places above 70°. Good for gluing boats or items that may be soaked because it is strong and waterproof. This glue must be mixed carefully. It has a dark color. |
| **White Polyvinyl** | Colorless, sets fast, and can be used at any temperature. Useful for small wood projects but is not water-resistant, so don't use on anything that is likely to get wet or damp. |

# SCREWDRIVERS

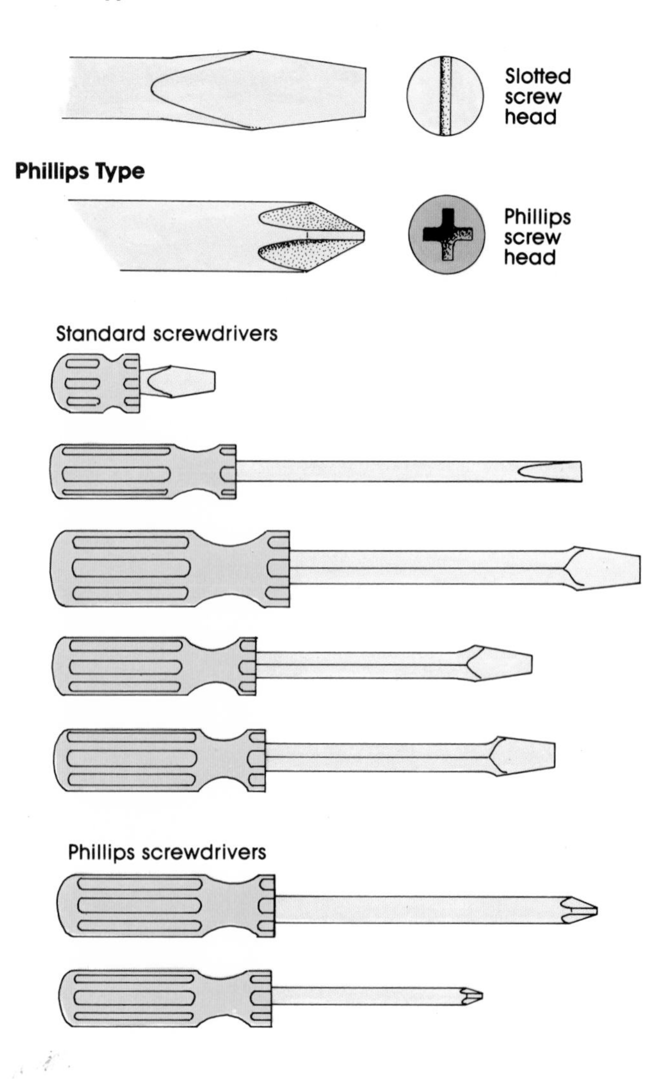


Because screws hold wood better than nails, every shop needs a good selection of screwdrivers. The screwdriver tip should always fit the screw head snugly. A tip that is too small will only slip and damage the screw head, making your work much more difficult. So you will want a variety of tip sizes, as well as both standard and Phillips tips. You will also need a selection of long and short screwdrivers, and several with straight tips and long shanks that allow you to drive deeply recessed screws.

Another type of screwdriver, relatively new to the market, has a ball head with a ratchet device built into it. The large ball handle provides much more torque and the ratchet device makes driving or removing screws faster and easier. It comes with an assortment of shanks and tips.

### Using a Screwdriver

Driving a screw into a piece of wood should be a relatively painless operation. When it isn't, you are doing something wrong.

First, drill a pilot hole for the screw (see pages 40–41). It should be slightly smaller than the screw so the threads

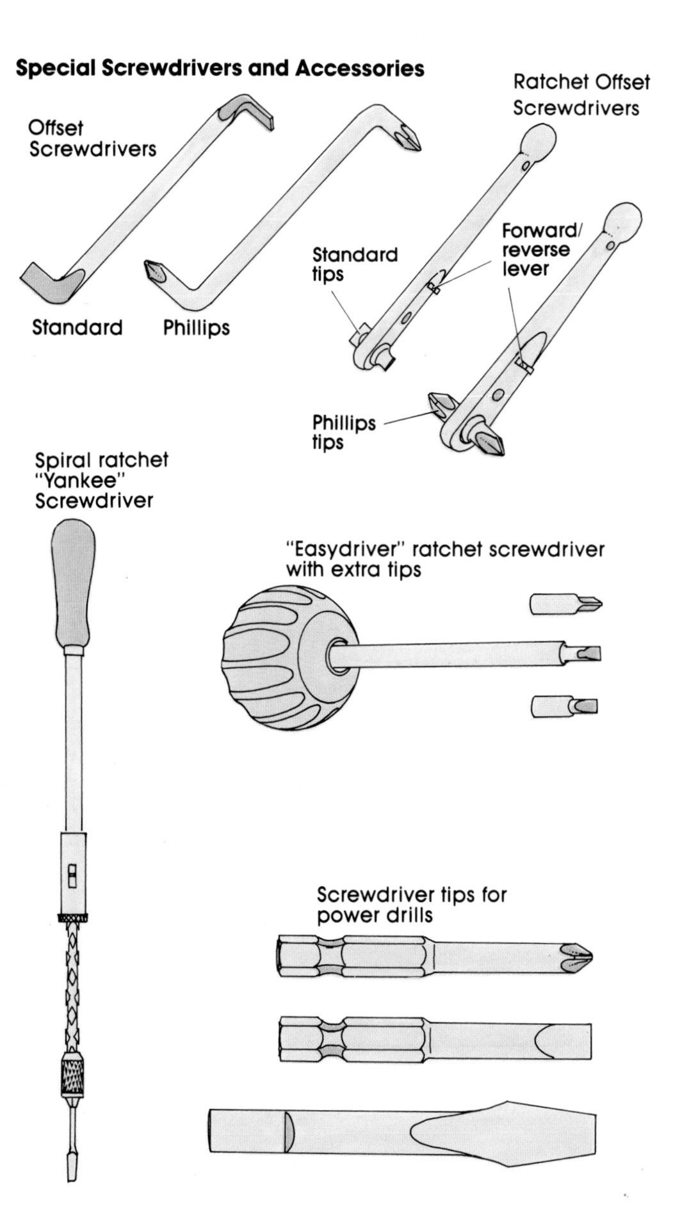


bite cleanly into the edges of the hole. Without the pilot hole, the screw has to crush its way into the wood. For small screws, you can punch a hole in the wood with an awl instead of drilling a hole.

Select a screwdriver with a tip that just fits the slot in the screw head (see the chart on Screwdriver Tips and Screw Sizes on this page). If the tip is much smaller than the slot, it will slip out and burr the slot until the screw can no longer be driven.

When driving screws with a power drill, use a slow speed on the drill so you can maintain control.

**Screwdriver Tips and Screw Sizes**

| Tip Width | Screw/Size |
|---|---|
| 1/4" | 6 to 8 |
| 5/16" | 8 to 12 |
| 3/8" | 12 to 16 |
| 7/16" | 16 to 20 |
| 1/2" | 20 to 24 |

Screws take longer to install than nails but have much greater holding power. They can also be removed if you need to take something apart.

Screw sizes are determined by their length in inches and by the size of their head, which is designated by a gauge number. Gauge numbers range from 2 to 16. Make sure that two-thirds of a screw's length is in the wood to which you are fastening.

## Wood Screws

The two most common types of screw heads are the slotted head and the cross-slot Phillips head. The heads themselves come in several different styles.

*Flathead screws* have flat heads that can be driven nearly flush with the surface of the wood. For a more finished appearance, they are either lightly countersunk so the head is completely flush, or countersunk and covered with wood putty.

The oval top on *ovalhead screws* provides more holding power for the screwdriver, making them easier to drive or remove. The tapered base under the head is normally countersunk.

*Roundhead screws* are general-purpose screws that are used where the high visibility of the head is not important. The flat base of the head provides a good support when fastened to thin material. Even greater support can be gained by using a washer under the screw head.

*Phillips screws* are similar in all aspects to flathead screws, except that the Phillips slot allows you to use greater turning power with less chance that the screwdriver will slip.

A *lag screw*, also known as a lag bolt, is a very large screw (up to 6 inches long) with a square head that is turned by a wrench. Lag screws are used to hold large pieces of lumber together.

*One-way screws* are slotted so they can only be tightened, not loosened. They are sometimes used to install protective shutters over vacation cabins. To remove them, the entire screw must be drilled out.

One end of a *hanger screw* is threaded to be driven into wood. The other end is threaded to accept a nut. It is useful for hanging metal on wood.

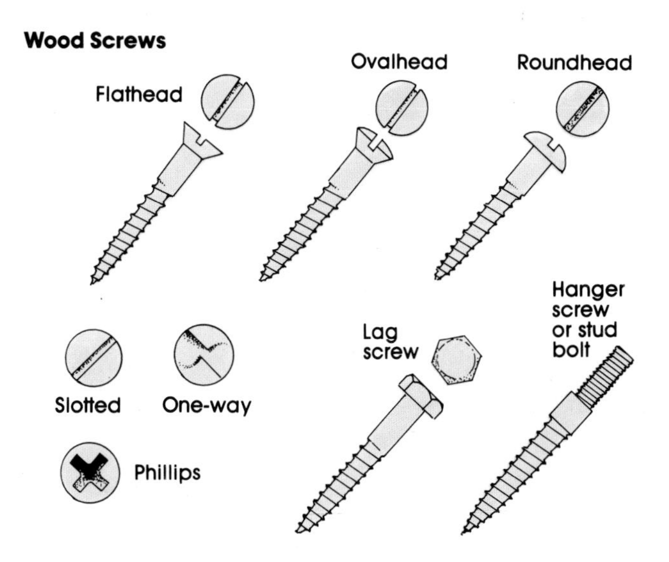

# BOLTS & NUTS

**Sheet Metal Screws**

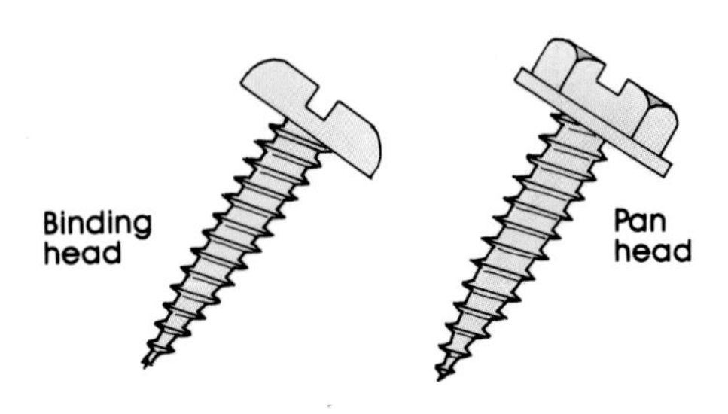


## Metal Screws

There are two basic types of screws used with metal: the *self-tapping* (or self-threading) screw, which cuts its own threads, or the *self-forming* screw, which turns its way around softer metal. The self-forming screw is commonly used to join two pieces of sheet metal, while the self-tapping screw is used for heavier-gauge metal.

*A note of caution:* when you work with aluminum, use aluminum screws (or nails), and when you work with galvanized metal, use galvanized screws or nails. Don't use galvanized screws on aluminum, or aluminum screws on galvanized metal. These opposing types of metal will establish a chemical reaction that causes the screws or nails to corrode.

## Bolts

Bolts are used where heavy-duty fastening is required. Rather than relying on threads sunk in wood to pull two pieces of wood together, as with screws, a bolt goes completely through the wood and then a nut is tightened on it to cinch the two pieces together. Bolts are widely used in construction: for example, for tieing two beams together or fastening a board to a masonry wall as a deck support.

Bolts are generally used with a washer, which provides a larger supporting surface and prevents the bolt or nut from digging into the wood as it is tightened.

Nuts come in several different styles and are usually sold in bins near the bolts so you can quickly match the nut and bolt.

A few basic bolts, washers, and nuts are illustrated on this page.

The *carriage bolt* is a good choice when you join together two large pieces of wood. A hexagonal shoulder right under the head that is slightly larger than the shank prevents the bolt from turning as the nut is tightened.

The *machine bolt* is used for either wood or metal. The head is either square or hexagonal for holding with a wrench while the nut is tightened. Machine bolts come with fine or coarse threads, with the coarse threads having slightly less holding power.

**Bolts**

Machine Bolts

Hexhead Squarehead

Carriage bolt

Stove Bolts

Flathead Roundhead Ovalhead

Cup hook

Eyebolt

Screw eye

Turnbuckle

**Molly Bolt**

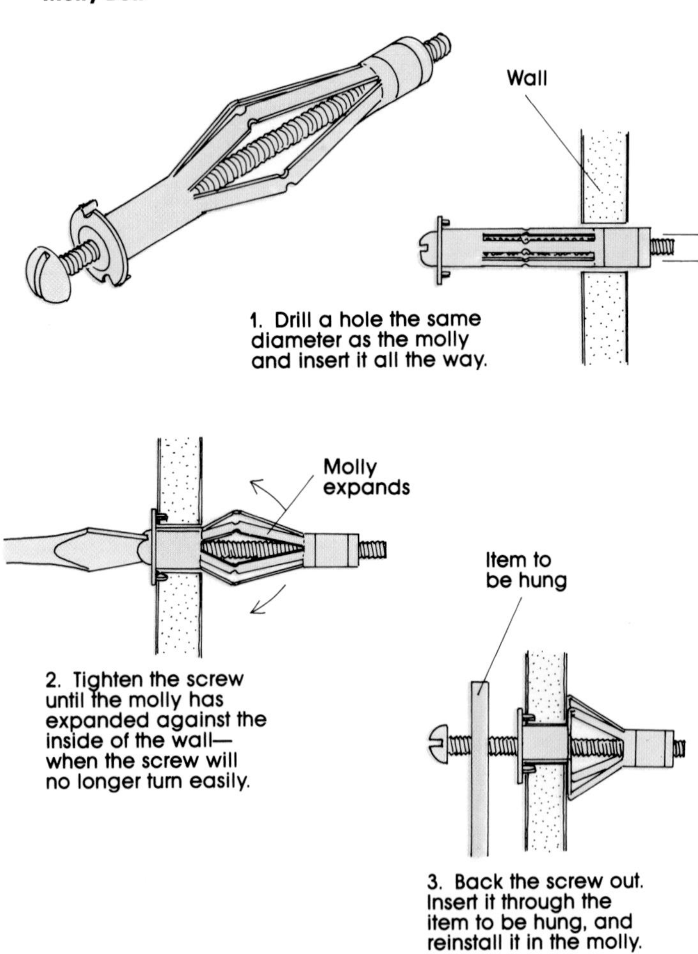


1. Drill a hole the same diameter as the molly and insert it all the way.

2. Tighten the screw until the molly has expanded against the inside of the wall—when the screw will no longer turn easily.

3. Back the screw out. Insert it through the item to be hung, and reinstall it in the molly.

The *stove bolt* is a utility bolt for holding light metals or wood together. The head, which can be flat, oval, or round, is slotted for a screwdriver.

The *toggle bolt* is used for fastening or hanging something in sheetrock where no support can be gained with a screw. The spring-activated clip at the end opens to provide broad support after being pushed through a hole drilled in the sheetrock.

The *Molly bolt* is also used for support in sheetrock or paneling. As the screw is tighetened, the metal legs flare out to grip the inside of the wall.

The *expansion bolt* is designed to fasten things to concrete or masonry walls. It consists of two parts: a lead or aluminum casing that fits in the hole drilled in the masonry, and the bolt. As the bolt is tightened, the casing expands to grip the edges of the hole.

The *eyebolt* is used to hold heavy hanging objects.

The *turnbuckle* is commonly used to tighten wire supports. Two eyebolts fit in the threaded sleeve, which has right-hand threads in one end and left-hand threads in the other. Just turn the sleeve to move the screws in or out.

## Nuts

The *hexagonal nut* is one of the most common varities. Its six sides can be gripped by either a standard wrench or a socket wrench. Socket wrenches do not fit well on square nuts.

The *square nut* is used interchangeably with the hex nut, and the lightweight flat square nut is commonly used with stove bolts.

The *cap nut* provides a more finished appearance where the nut is visible.

The *wing nut* is easy to tighten or loosen by hand and is commonly used with small stove bolts.

The *knurled nut* has knurled edges that are easy to grip with fingers or pliers.

## Washers

There are basically two types of washers: flat washers provide more support under bolts and nuts, and locking washers keep nuts from working loose.

*Flat washers* come in a great variety of sizes and thicknesses, depending on the need. For maximum support, use one under the head of a bolt and one under the nut.

*Locking washers* are used to keep the bolt from working itself loose. Spring lock washers keep tension against the nut, while the internal or external tooth washers bite into the wood or the nut.

**Toggle Bolt**

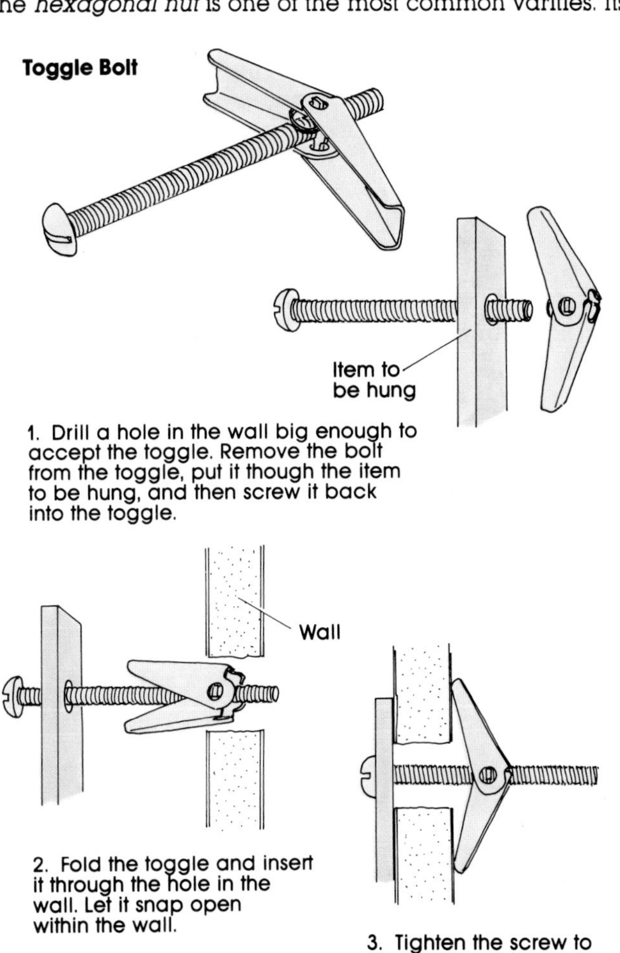


1. Drill a hole in the wall big enough to accept the toggle. Remove the bolt from the toggle, put it though the item to be hung, and then screw it back into the toggle.

2. Fold the toggle and insert it through the hole in the wall. Let it snap open within the wall.

3. Tighten the screw to pull the toggle against the inside of the wall.

**Expansion Shield**

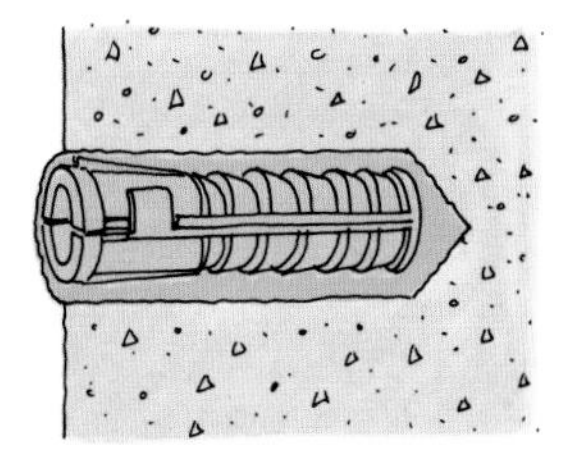

1. Drill a hole in the masonry large enough that the shield slides in easily all the way.

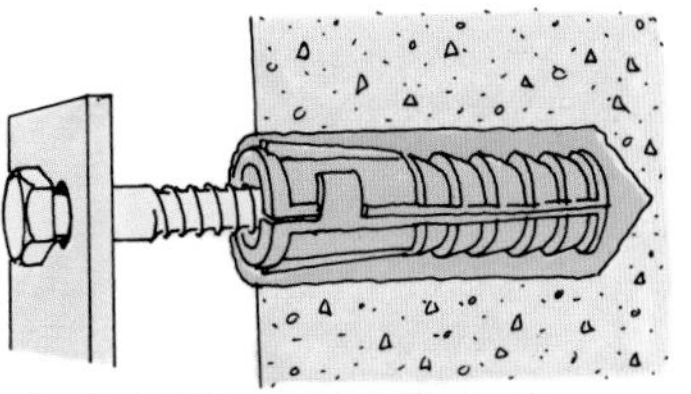

2. Put a lag screw through the item to be hung and into the shield.

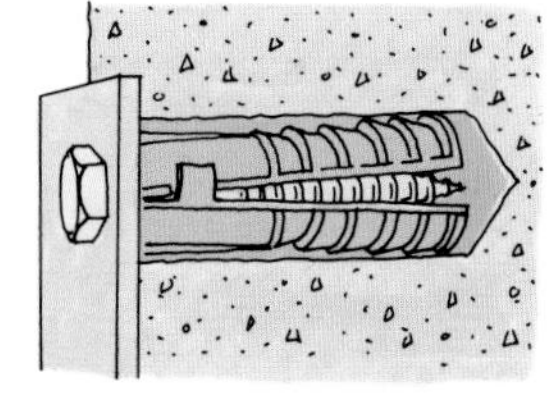

3. Tighten the bolt and the shield will expand and hold it securely.

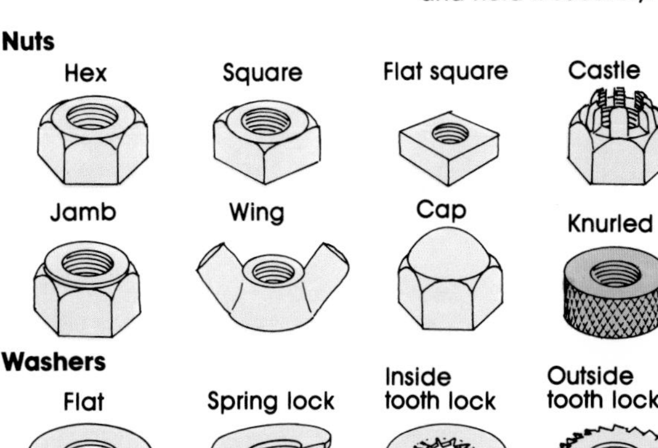

# WRENCHES & PLIERS

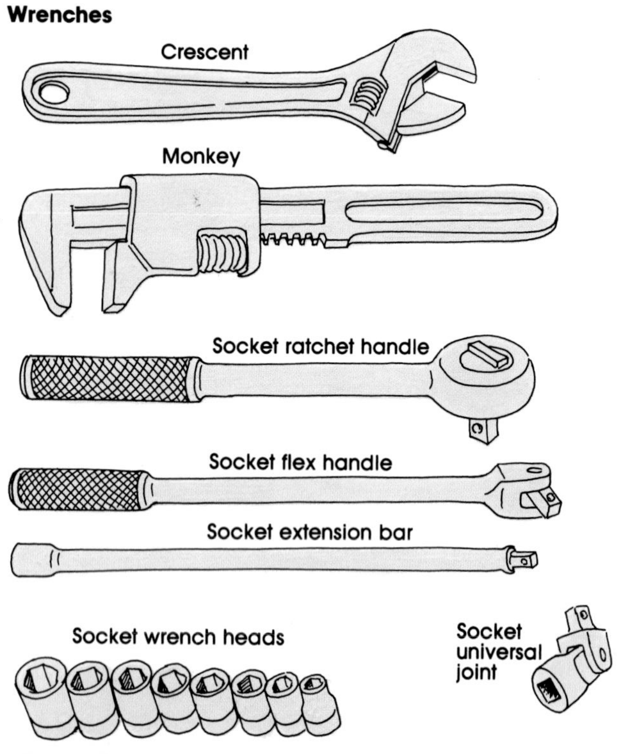


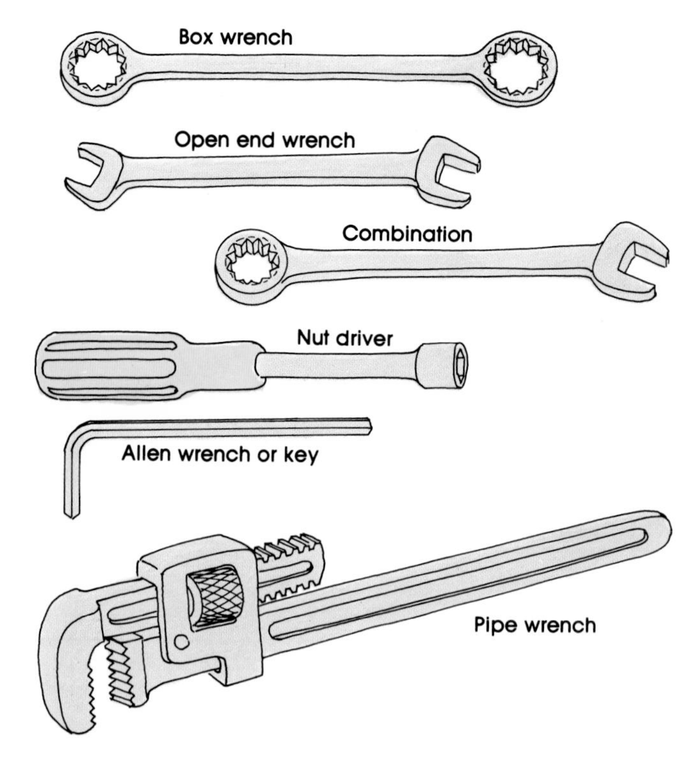


## Wrenches

No workshop is complete without an assortment of wrenches. They are the workhorses for tightening or loosening nuts and bolts, turning lag screws, or working on pipes.

The standard *adjustable wrench,* available with 8- or 12-inch jaw openings, should be within easy reach around a workshop. Two wrenches are usually required: one to hold and the other to tighten.

The *socket wrench* is used with a selection of sockets that commonly range from ⅜ inch to 1 inch. Smaller and larger sizes are available. It works faster and holds a bolt or nut more firmly than an adjustable wrench.

*Pipe wrenches* should be standard for the home workshop. Again, a pair of these is required, one to turn and the other to hold. In addition to pipe, this wrench can be used for loosening a big nut or bolt that is badly burred and refuses to turn.

Other important types of wrenches for turning nuts and bolts are the *box wrench,* the *open end wrench,* and the *combination wrench.* These come in specific sizes, commonly ranging from ¼ inch to 2 inches. The openings at each end of the wrench are usually different sizes.

The *nut driver* is handy for small nuts and bolts. It resembles a screwdriver, only with a socket that fits over the nut on the end. It comes in different sizes to fit different nuts.

An *Allen wrench* is used to turn an Allen head screw.

## Pliers

Like your fingers, pliers are used to grip things, but they are hundreds of times stronger than a human hand. A basic selection of pliers is illustrated on this page. Having the right pliers at the right time will save you much frustration.

If pliers won't hold properly, but are slipping and burring the piece, you should be using a wrench instead. For example, beginning carpenters often turn nuts with pliers instead of a wrench.

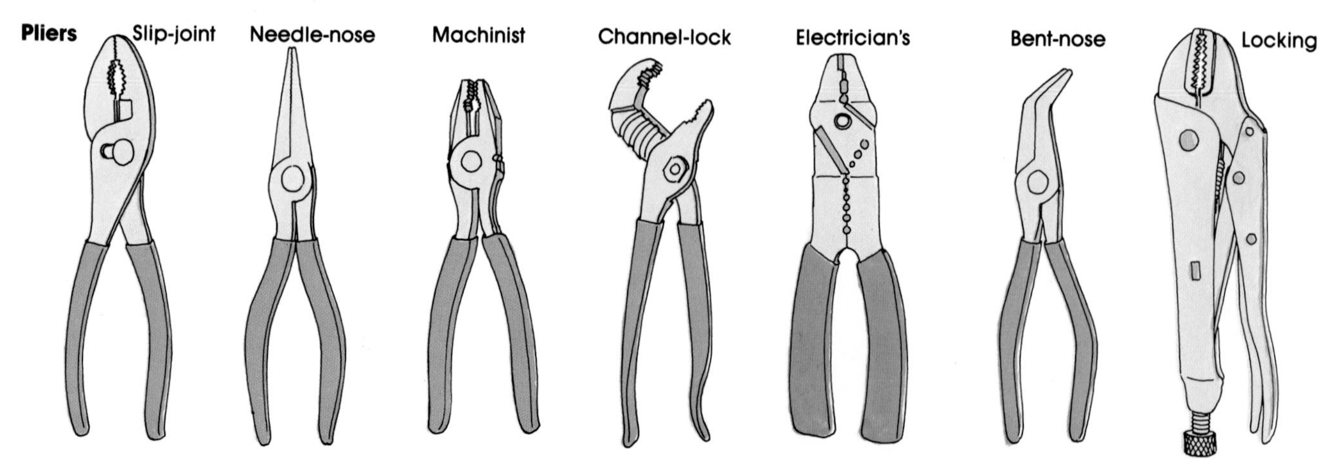

# FILES & RASPS

There are more than 3,000 kinds of files and rasps, but you can get by with a few of the basic types. You need a selection of metal files as well as wood files and rasps.

Files can be used on metal or wood, but rasps are used for wood only. Files have ridges, whereas rasps have individual triangular teeth. Rasps do rougher but much faster work on wood than files.

Files have two types of ridge configurations: the *single-cut* has parallel ridges or serrations running across the file and is used for fine filing or shaping; and the *double-cut* has ridges running at opposite angles to each other in a checkerboard pattern and is used for coarser work.

Files range generally in size from 4 to 14 inches long. The larger the file, the more space there will be between the serrations, and thus the coarser the cut it will make.

Files are designated by the coarseness of their cut. The four basic types of files are coarse, bastard, second, and smooth. Each type has progressively smaller spaces between the serrations, thus providing progressively smoother filing action.

Files and rasps come without handles, which must be purchased separately. The tang, or pointed end of the file, is forced into the handle. Use a handle; it's safer and gives you more control. Also, keep a stiff wire brush handy to clean your files. They don't work as well when the serrations or teeth become clogged with metal or wood scrapings.

Rasps typically come with one flat side for filing wood smooth, and one rounded side for shaping wood. A good choice for the workshop is a wood rasp with four sections, each with a different coarseness.

Another woodworking device on the market is the Sureform tool. It is shaped somewhat like a plane, which makes it easy to hold, and the rasps that screw into the base are interchangeable, depending on whether you want coarse or fine results.

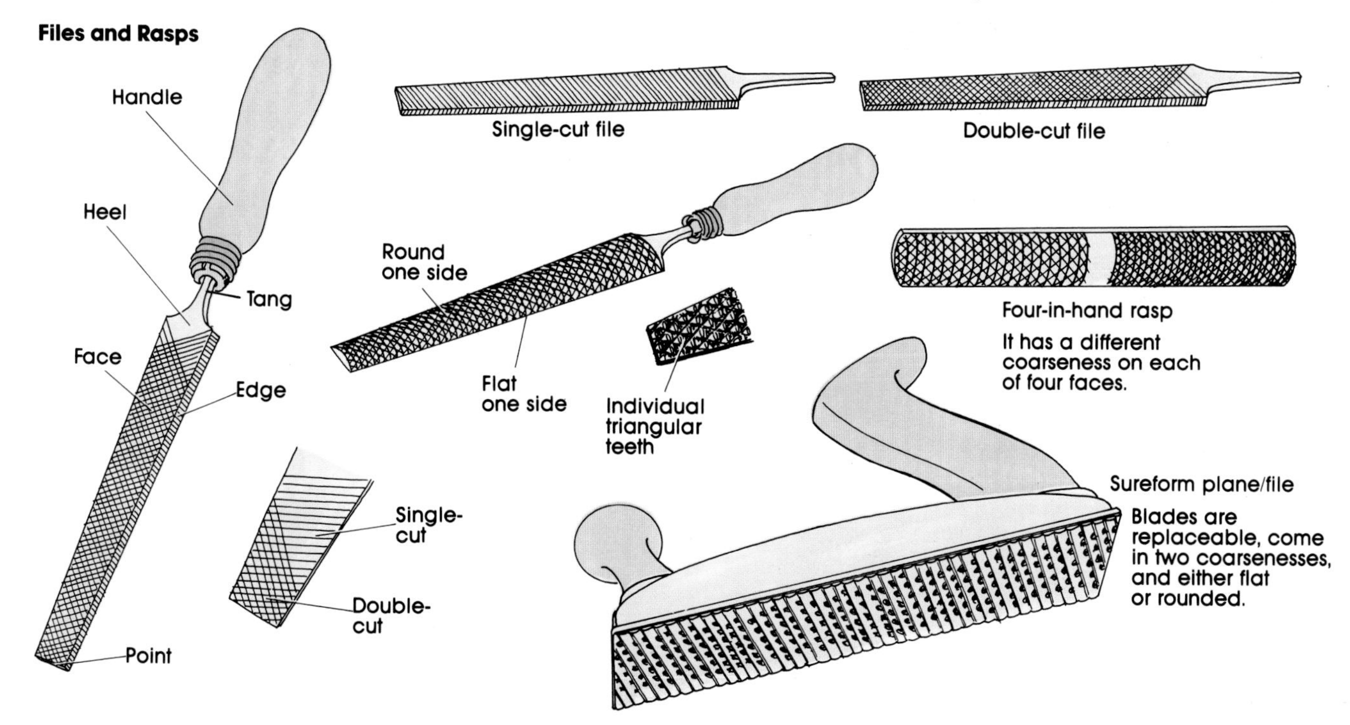


# CHISELS & GOUGES

**Chisels**

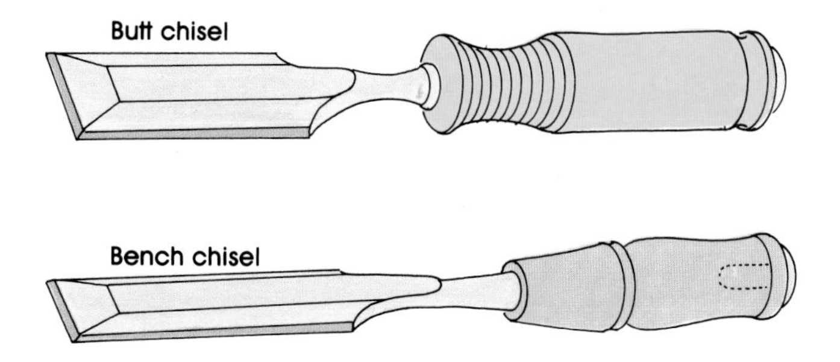


Chisels and gouges come in a tremendous variety of shapes and styles, depending on whether they are designed for use by woodcarvers, carpenters, electricians, lathe operators, or sculptors. All are used for trimming and shaping.

Chisels come with wood or plastic handles. For rough carpentry work, plastic handles are preferred because of their greater strength. The handle end should have a steel cap on it to take the vigorous pounding it will receive.

Some of the basic types of chisels used in carpentry work are illustrated on this page.

The *butt chisel* has a short blade, about 3 inches long, and ranges in width from ¼ to 1 inch. The blade is fairly thick and is designed for heavy work, such as trimming rough boards for a smooth fit or cutting out notches. Many carpenters keep a 1-inch butt chisel in their tool belt for rough trimming on job sites.

*Bench chisels* are used around the workbench when precision and fine cutting are required, as in cabinetmaking. There are three types of bench chisels. The *paring chisel* is the lightest, with the edge beveled to about 15 degrees. It is used for delicate work. The *firmer chisel,* for medium work, has an edge beveled to about 20 degrees.

The *framing chisel* is designed to remove larger amounts of wood and has an edge beveled to about 25 degrees.

*Cold chisels* are made from hardened steel and are used to cut metal. They are useful around a shop for tasks such as cutting bolts when the nut is so rusted it cannot be turned.

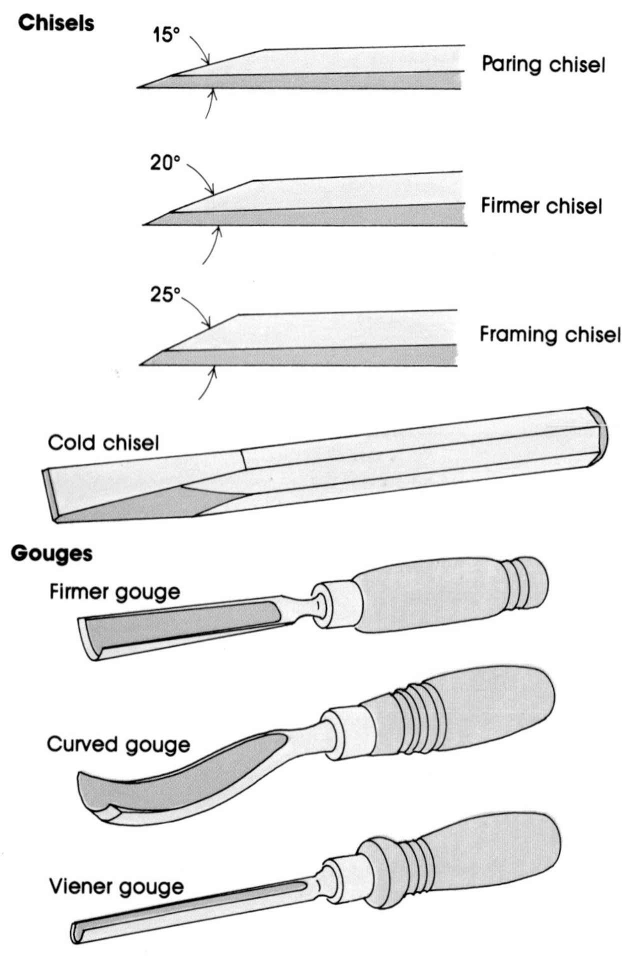


Gouges are similar to chisels except that the blades are curved for cutting grooves and hollows. They are mainly used for cabinetwork.

## Using Chisels

Chisels can be powered by the hands alone when working with softwood or removing fine amounts of wood. When you work with hardwood or are removing large amounts of wood, power the chisel with a hammer or mallet. If you are right-handed, hold the hammer in the right hand. Always use a mallet on chisels with wooden handles; use a hammer only on plastic-handled chisels that have a metal cap on the top.

You should normally cut with the bevel down. The exception to this rule is when you are removing wood from a notch; then keep the bevel up.

Always cut with the grain. Working against the grain will result in splitting off large pieces.

When you cut out grooves or notches, use a chisel slightly smaller than the notch. If you use a 1-inch chisel in a 1-inch groove, for instance, you will tear the sides.

# PLANES

**Parts of a Plane**

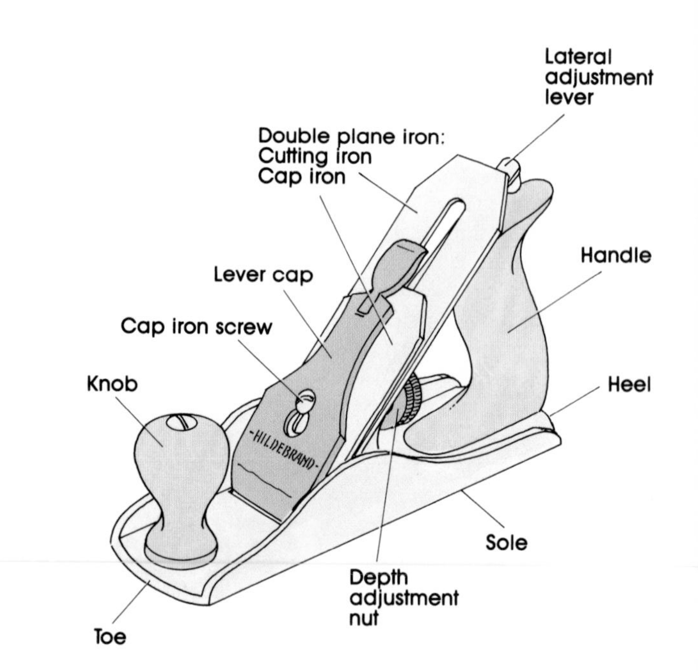


Keep your fingers clear of the cutting edge.

There are three basic types of planes: the small block plane, the workhorse jack plane, and the long and elegant jointer plane. While all three are designed to remove wood in thin, controlled shavings, each one does it a little differently.

The *block plane,* small enough to use with one hand, is good for rounding corners, beveling, or removing small amounts of wood.

The *jack plane,* usually about 14 inches long and 2 inches wide, provides a smoother and more even cut than the block plane because of its added length. It is used for general planing.

The *jointer plane,* up to two feet long, provides the smoothest, straightest cut of all. This is what you need if you want to fit boards together with almost invisible seams, such as for a tabletop.

A plane is probably the most complex nonelectrical tool in your workshop. You should know how to take it apart, put it back together, and make the necessary fine adjustment. The following discussion will help you do this.

The key elements of the plane are the double plane iron, the level cap, the depth adjustment screw, and the lateral adjustment lever. All must be carefully set for the plane to function properly.

**The double plane iron.** The plane iron is the beveled cutting piece. The plane iron cap, with its curved end, rolls the wood shaving out to keep the blade from clogging.

To adjust these two pieces, slide the cap down just 1/16 inch from the plane iron's cutting edge. Make sure it is square and then tighten the cap iron screw.

**The lever cap.** Now slip the double plane iron into the body of the plane, with the bevel edge down and the plane iron cap on top. Next, place the lever cap over the

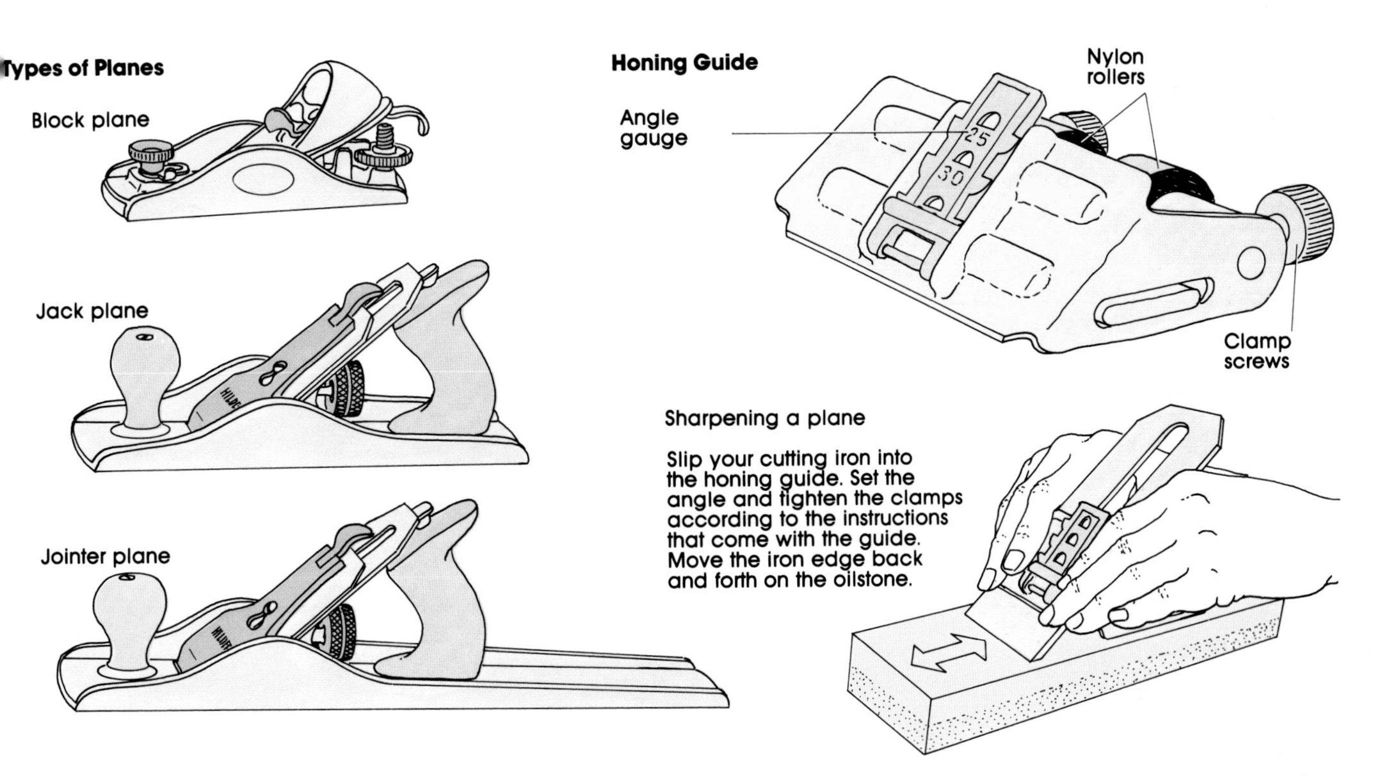


plane iron cap. The cam at the end of the lever cap should be up. Hook the lever cap over the lever cap screw and then snap the cam lever down.

If the plane is properly adjusted at this point, the cam lever will snap down firmly and the lever cap and double plane iron will be firmly in place. If it is loose, remove the lever cap and tighten the lever cap screw by a quarter turn. If the cam lever is too difficult to depress, loosen the screw slightly.

**The depth adjustment screw.** Turn the plane upside down and sight down the face of the plane at the cutting edge. For the jack plane, you want the blade to protrude about 1/16 inch. For jointer planes, which do finer work, let only a hair's thickness show. To make these adjustments, turn the depth adjustment screw counterclockwise to extend the blade and clockwise to draw it back up.

**The lateral adjusting lever.** While the blade may be extending to the proper depth, it may be cocked to one side or the other. Again, sighting along the face, make the necessary adjustments. Push the lever to the left to extend the right corner of the blade; push the lever to the right to extend the left corner of the blade. If the blade remains severely angled even with the adjusting lever centered, take the plane apart again and check that you put the double plane iron together straight.

## Planing Techniques

The first rule for using a plane is to work with the grain of the wood. Otherwise, you will just tear the wood apart and clog the plane.

If you find it difficult to operate the plane and you have a thick curl of wood coming out, adjust the plane for a finer cut (see above).

The most common error in using a plane is the application of pressure in the wrong manner, which results in rounding of the wood surface.

The proper way to plane is as follows. Place the plane squarely on the surface. As you begin to move forward, maintain pressure on the toe while pushing on the handle. As the toe reaches the end of the board, relax the pressure there and press down on the handle. Don't try to cut too much at once; keep the shavings paper thin.

If you start planing with pressure on the heel, under the handle, and then push down on the toe at the end, you will tilt the plane and bevel the ends of the board. If you find that despite your best efforts the board seems to be hollowing out in the middle, you should use a longer plane.

When you are working on a broad, flat board, test your work after a few passes with the plane by placing a square across the board and sighting down it. You know you have depressions wherever you see light coming under the square.

When you are planing the edge of a board or door, it's important to keep the plane flat and not to apply extra pressure to one side or the other. Again, check your work with a square to be sure you are not beveling one side.

When you are working on the butt end of a board or door, plane from one side to the middle and then from the other side to the middle. Trying to run across the end in one pass may cause the far side to split. Another trick is to clamp a piece of scrap to the far side, flush with the top, to prevent the end from splitting off.

Because the blade of a plane is so sharp, it tends to dull and nick rather easily. Keep it sharp by regularly running the blade over a whetstone with the bevel down and perfectly flat. You can buy a special tool to hold the plane in position while you whet it. (See page 62 for tips on sharpening your tools.)

When you store the plane, always put it on its side in order to avoid dulling or damaging the blade.

# SANDERS

Sanding is an essential part of all finish carpentry. You can plane or file wood smooth, but only expert sanding can give the final product a professional appearance.

Sanding has traditionally been done by hand, a method that still gives fine results. But power sanders do the job just as well and considerably faster. An orbital sander, for instance, makes 9,000 strokes a minute; by working hard, you might make 300 strokes a minute.

If you will be doing much cabinetmaking at all—even one basic table—either a belt sander or an orbital sander is a good investment.

## The Belt Sander

The size of a belt sander is designated by the size of the abrasive belt it takes. The belts range from 2 to 4½ inches wide and from 21 to 27 inches long. The bigger the sander, the faster it does the job. For all-purpose work, choose a heavy-duty sander that takes a 3-by-24-inch belt.

To install the belt, unplug the sander and turn it on its side. Most models have a tension release lever that is pulled up. Slip on the belt and push the lever back down. Now hold the sander in the air with one hand and turn it on. You'll notice that the abrasive belt begins drifting to one side or the other on the front roller. Turn the belt-aligning screw on the front left side slightly one way or the other until the belt tracks properly.

### Using the Belt Sander

Always keep a belt sander moving while sanding. If you hold it in one place, it will quickly cut grooves in the wood. Turn it on and lower it gently onto the wood, with the heel coming down slightly before the entire belt touches. Don't press down on the sander; its own weight is sufficient pressure. Move the sander back and forth with the grain as you slowly work your way down the board. At the end, let the sander ride more than halfway off the end of the board, but don't let the nose dip or you will round off the butt.

For faster surfacing, turn the sander at an angle to the board while still moving it back and forth with the grain. Angling it takes the top off the grain faster. Finish up with the sander held in line with the grain.

If you are sanding two pieces of wood that are joined with their grains running at opposing angles, cover one with tape while sanding the other. Sanding against the grain leaves marks you don't want.

To check your work for grooves, place your eye almost level with the wood and look toward the light source.

## The Orbital Sander

The orbital sander, also called the finish sander, works more slowly but produces finer results than the belt sander. The orbital sander moves in a tight circle, each stroke moving only 3/16 to 1/4 inch. Because it is moving in all directions, you don't have to move the sander with the grain.

Care must be taken to fit the sandpaper tightly over the sanding pad on an orbital sander. The paper is held by clamps at each end.

Turn the machine on before you lower it onto the wood. The weight of the sander and your hands is enough. Pressing down more will slow the sander down and overheat the motor.

**Parts of a Belt Sander**

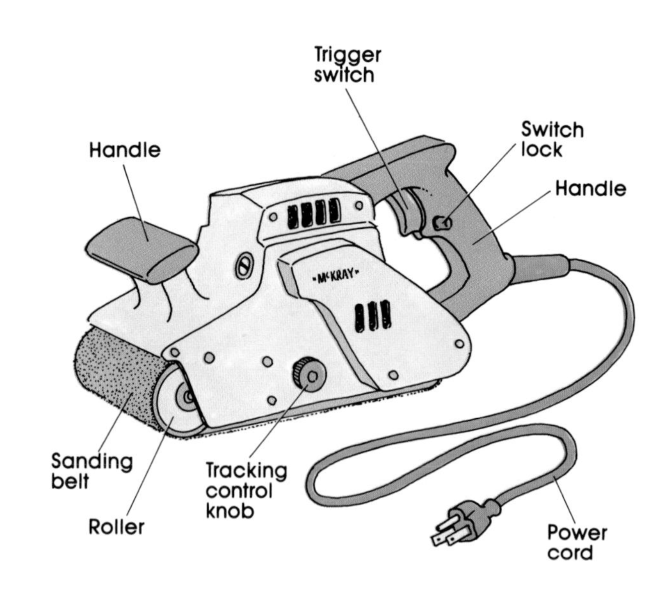


**Parts of an Orbital Sander**

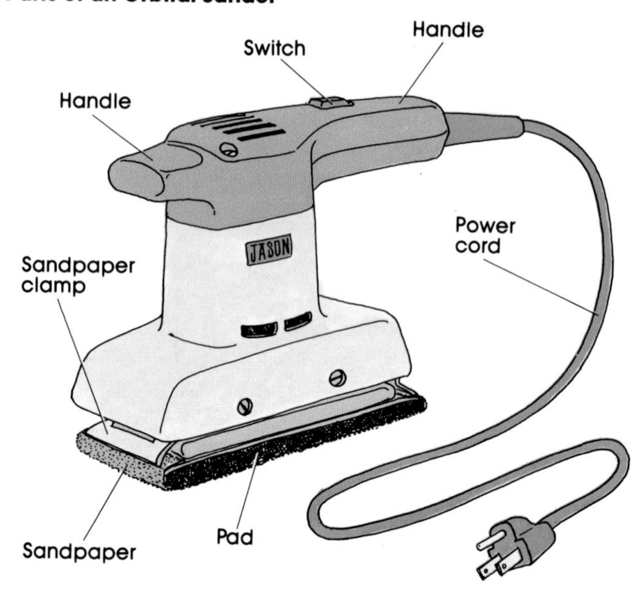


Before you give the wood its final sanding, wipe it down with a damp sponge or rag. This removes sanding dust from between the grain lines. The final sanding will now smoothly level the raised grain.

## Care and Cleaning of Sanders

Both belt and orbital sanders tend to collect dust in the motor. If you have an air compressor, blow out your sander regularly. Otherwise, use a brush to clear out the dust.

Some sanders have a dust filter on the front. You can remove the filter and tap out the dust periodically, but don't use an air hose on the filter or you may break the mesh.

Check your sander regularly for loose screws and damage to the pad. The pad can be cut by running over nail or screw heads in the wood.

## Sanding by Hand

In all sanding, it's essential that you sand with the grain. Sanding across it or at random will cut the tops of the grain and may do irreparable harm. Those tiny cuts will soak up stain faster than the other wood and leave you with a patchwork effect.

When you are working on flat surfaces, wrap a piece of sandpaper around a block of 2 by 4 cut to fit comfortably in your hand. This provides a flat, uniform surface that your fingers can't give. Commercial sandpaper blocks are also available.

## Sandpaper

A stamp on the back of a piece of sandpaper tells you the weight of the paper backing, the type of abrasive, and the grit number. Each of these variables is described below.

**Paper backing** comes in five weights, each designated by a letter. The letter A represents the lightest weight, which is used for finish work. B, C, and D represent progressively heavier backing, and E is the heaviest weight, used primarily for commercial work.

**Abrasives** vary widely, depending mainly on the material they are made from. Sandpaper is usually designated as either close-coated or open-coated. The abrasives on open-coated sandpaper are spread farther apart, which is preferable for working on softwoods such as pine. This is because pine tends to be resinous and will quickly gum up the more densely packed close-coated sandpaper. Close-coated sandpaper is preferable for finer work, such as on hardwood.

Commonly used abrasive materials include the following:

*Flint* is one of the poorest quality abrasives because of its softness. Sometimes it is suitable for rough sanding or removing paint.

*Garnet*, a semiprecious stone, makes an excellent abrasive for finish work.

*Aluminum oxide* is a synthetic garnet that makes a tougher abrasive, but it is also rather coarse. Use it for the first sanding to remove excess roughness.

*Emery* is harder yet, is generally applied to a cloth backing, and is excellent for giving metals a fine polish or cleaning.

*Crocus* cloth is similar to emery and is widely used for sanding metals.

*Silicon carbide* is second only to diamond in hardness but is brittle and breaks easily if too much pressure is applied. It is good for sanding soft metals or plastic.

**Grit numbers** are determined by passing the abrasive through a 1-inch square mesh and commonly range from 12 to 600. The bigger the number, the finer the sandpaper.

### Using Sheets of Sandpaper

Before you begin to work with a sheet of sandpaper, place it on the edge of your workbench, abrasive side up, and pull it around and under the edge of the bench. This causes numerous fractures in the glue backing and greatly reduces the chance that the backing will peel away in one strip. To tear sandpaper, fold it with the abrasive side in, crease it, and then make the tear.

### Sandpaper Belts

Information about belt sandpaper is stamped inside the belt. The abrasives for belt sandpaper have different designations, ranging from 2½ for very coarse down to 4/0 for very fine (see the illustration on this page). Sandpaper belts always have a cloth backing. The belt also has an arrow stamped on the inside that indicates the direction the belt must turn.

Your sander will be stamped with an arrow indicating the direction of rotation. Make sure you install the belt with the arrow matching the machine's rotation direction.

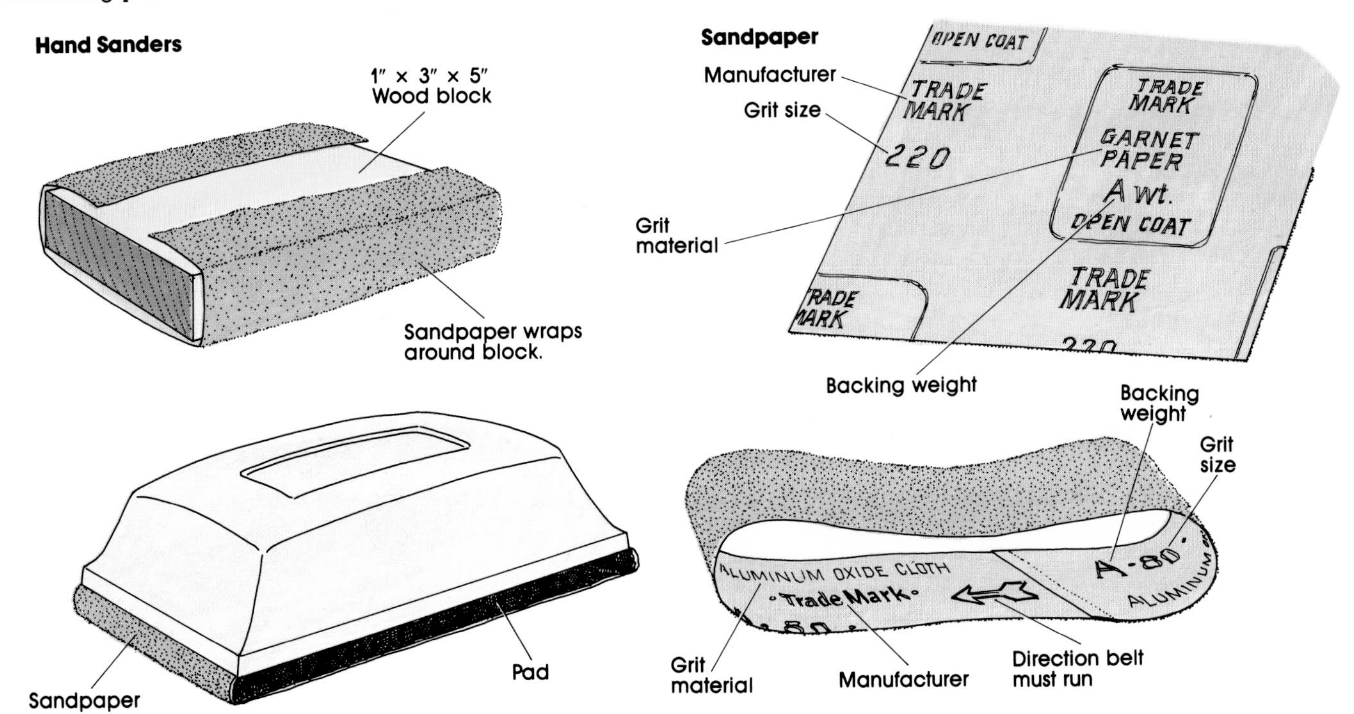

# MEASURING TOOLS

Measuring Tapes

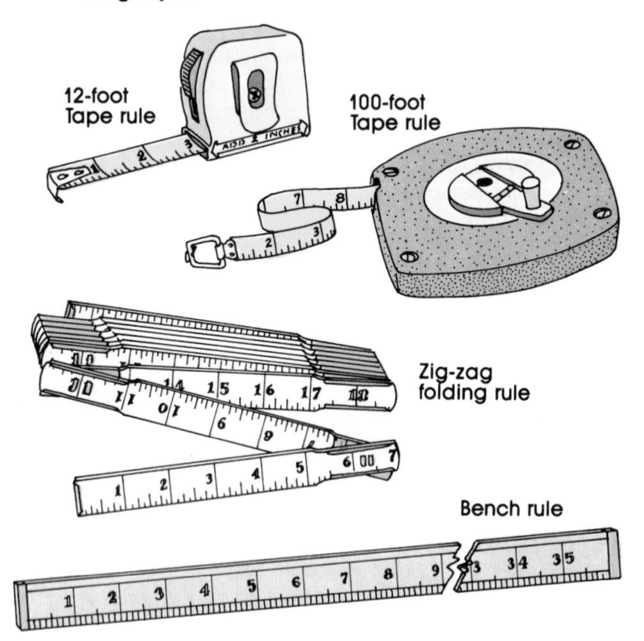


## Tape Measures and Rulers

When measuring a board, you should always mark the point with an inverted V. The point of the V should be right on the fraction of an inch to which you measured. Don't mark with just a dot; it's too hard to locate. And don't mark with just a little line; the line won't be straight and you may forget whether you are supposed to measure to its top or its bottom.

**The retractable steel tape** is the standard measuring tool for the rough carpenter. It has a hook on one end to catch the butt of a board while you measure several feet away from it. The hook is calibrated to move in or out by an amount equal to its own thickness so that you get consistently accurate measurements for either outside or inside dimensions.

These tapes range from 6 to 25 feet long, with the 12-foot tape being most practical for the carpenter. There are usually too many boards 8 feet or more in length to warrant using a shorter tape. A good choice for a second tape is the 100-foot steel tape, which is useful for measuring longer distances and especially for measuring the diagonals of foundations and stud walls to check that they are square.

**The bench rule,** similar to a school desk ruler, is so named because it is kept around a workbench. Because bench rules are calibrated down to 1⁄16 and even 1⁄32 inch, they are most often used by cabinetmakers. They are made of either steel or hardwood. The wood rule is generally thicker than the steel one and must be placed on edge so the calibrations will be flush with the wood to be marked.

**The zig-zag folding rule** used to be standard in any workshop but is now less frequently seen, although it is still an excellent ruler. It is longer than a standard bench rule and remains rigid when it is fully extended, which makes for easier handling than a steel tape. The better-quality folding rules have a brass extension rod at one end that slides in or out for precise inside measurements.

## Levels

There are many different types of levels on the market, but the principle is always the same: a bubble in a glass vial of alcohol or ether rises to the top of the vial to indicate a level surface. For years, most levels had a slightly curved vial on each of two edges so that the bubble would rise to the top regardless of which edge you put down. Nowadays levels usually have a single vial with a slight bulge in the center for the bubble. This type of vial will work whether it is used on edge or laid flat.

A level takes a lot of knocking about and is inevitably going to be dropped, so buy a good one. You can usually judge quality by the price. High-quality levels can be made from either wood or metal.

For most jobs, the standard two-foot *carpenter's level* is a good choice. It has a vial in the center to check the horizontal and a vial at either end to check the vertical. It can be made of wood, aluminum, or magnesium.

A *mason's level* is much longer—up to 6 feet. The additional length is needed to cover several masonry blocks at once so that it doesn't give a false reading just because one block is slightly out of level.

If you are laying blocks for a foundation and don't want to buy a long level, you can get the same effect by laying a straight 2 by 4 on the blocks and then placing your shorter carpenter's level on it.

The *line level* is only a couple of inches long and has hooks at each end so that it can be hung on a piece of string. It's useful for giving close approximations on long reaches—for example, when you are laying out the strings for a foundation.

A *torpedo level* is usually less than a foot long and is something of a compromise between the carpenter's level and the line level. Its short length allows you to work in tight places and still get an accurate reading.

Types of Levels

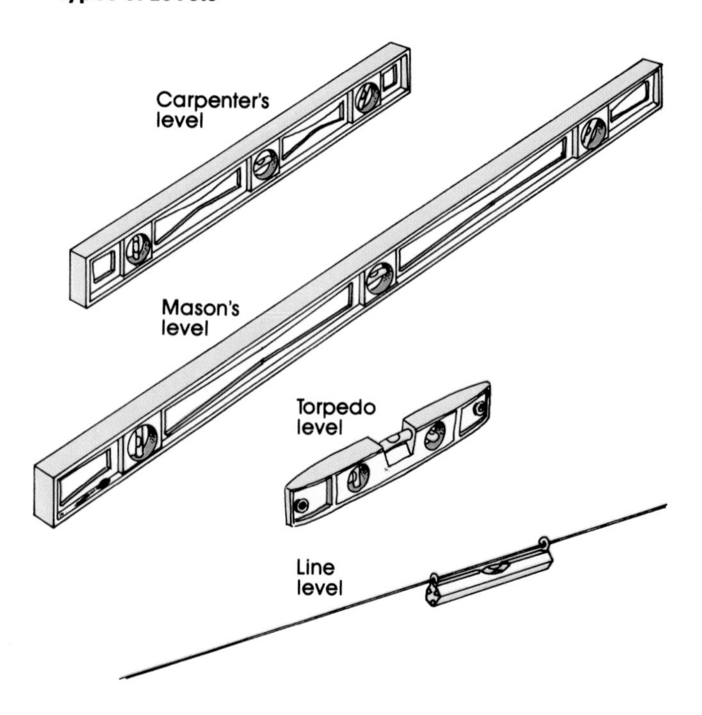

## Squares

### Types of Squares

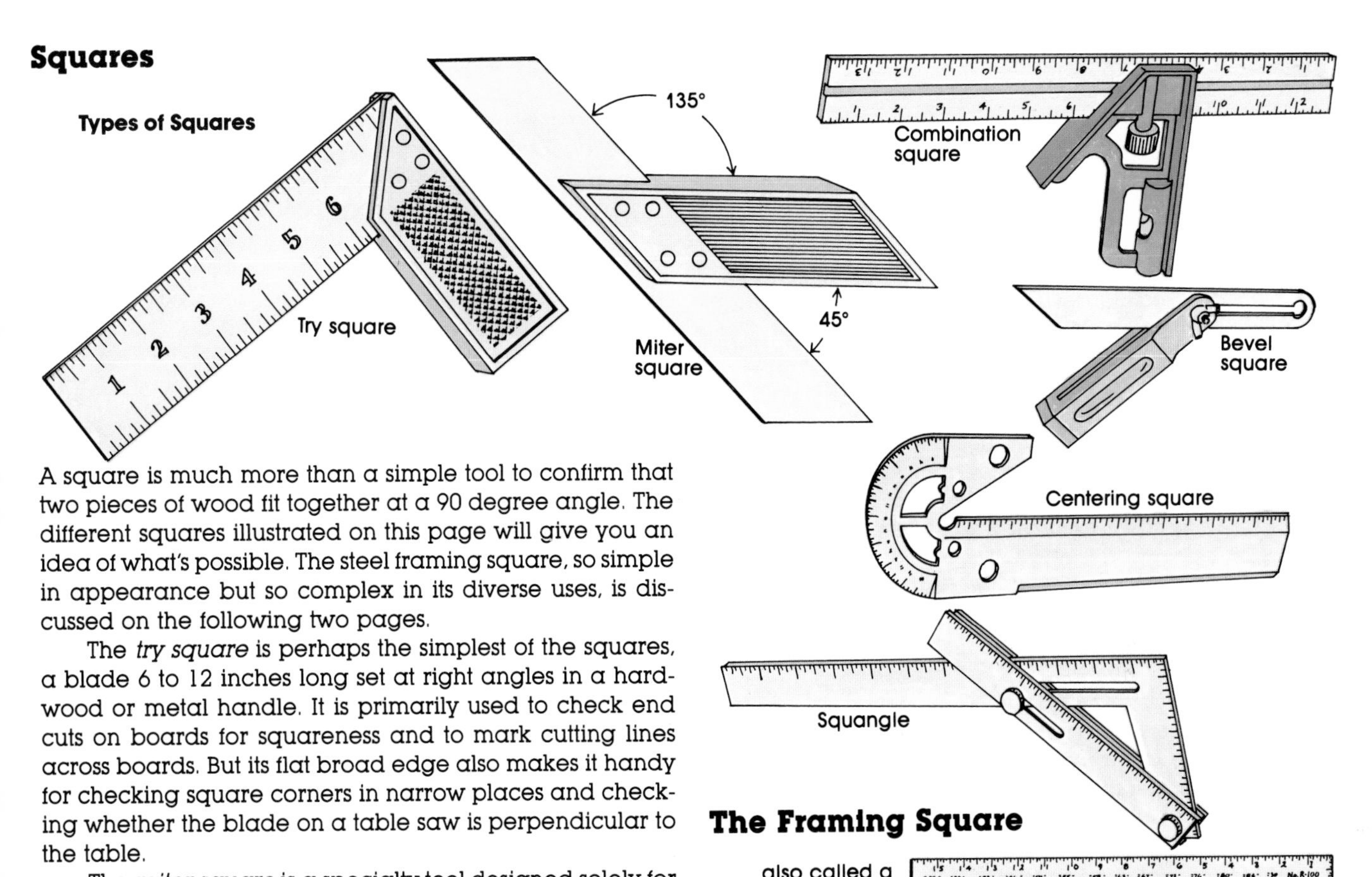


A square is much more than a simple tool to confirm that two pieces of wood fit together at a 90 degree angle. The different squares illustrated on this page will give you an idea of what's possible. The steel framing square, so simple in appearance but so complex in its diverse uses, is discussed on the following two pages.

The *try square* is perhaps the simplest of the squares, a blade 6 to 12 inches long set at right angles in a hardwood or metal handle. It is primarily used to check end cuts on boards for squareness and to mark cutting lines across boards. But its flat broad edge also makes it handy for checking square corners in narrow places and checking whether the blade on a table saw is perpendicular to the table.

The *miter square* is a specialty tool designed solely for marking 45 degree angles or bevels, or for a 135 degree angle on the opposite side.

The *sliding bevel square* (or bevel square as it's commonly called) adjusts to any angle and is held tight by a wing nut. When you don't know the actual measurement of an angle in degrees, you can use this square to match it to the board you are going to cut. If you do know the exact degree of an angle, you can place the bevel square on a protractor and preset the square to the desired degree.

The *combination square* is the workhorse tool for the carpenter. It can be used quickly to check 90 and 45 degree angles, contains a small level, and can also be used as a ruler. The illustrations demonstrate its diversity.

The *squangle* is a relatively new tool that takes a lot of the mystery out of marking rafters for the desired pitch (angle) of a roof. It is marked with a scale and quickly adjusts to whatever pitch you require. It also contains a scale for calculating rafter lengths or hip and valley rafter lengths, plus a degree marking for cutting hip and valley rafters. Instructions are contained in the packaging. (See pages 96–98 for a discussion of framing.)

The *speed square* is another tool that makes it easier to calculate angle cuts for rafters, including hip and valley rafters. It comes with instructions on how to determine the rise of a roof and with tables for determining lengths of rafters.

The *centering square* is an inexpensive specialty tool that helps you locate the center of a circle quickly and then doubles as a protractor to measure angles. Instructions are included on the packaging.

## The Framing Square

also called a Carpenter's square

The framing square contains a lot of information. It enables you to calculate the rise and pitch of a roof plus all the angled cuts necessary for common, valley, hip, and jack rafters. It tells you how to calculate board feet, lay out an octagon, or calculate the length of a 45 degree angle brace. It is also an essential tool for laying out the cuts to be made in stair stringers. While not all the information is commonly used by carpenters, anyone planning to build the simplest frame house should know how to use the square to lay out the rafter and stair stringer cuts.

The framing square is made from a single piece of steel bent to form a right angle. The *tongue* is usually 16 inches long and 1½ inches wide. The *blade* is 24 inches long and 2 inches wide. The point of the right angle is the *heel.*

There are two distinct sides to the framing square: the face, which has the manufacturer's name on it, and the back. The edges on both sides are calibrated in inches—but with an important difference. The face of the blade is calibrated in 1/16-inch fractions on the outside and 1/8-inch fractions on the inside. The back of the blade is calibrated in 1/12-inch fractions on the outside and 1/16-inch fractions on the inside. The back of the tongue is calibrated in 1/12-inch fractions on the outside and 1/10-inch fractions on the inside. There is also a scale on the back near the heel that is calibrated in 1/100-inch fractions.

The framing square also includes a scale for calcu-

# ADDITIONAL TOOLS

lating board feet, but it only works for boards that are 1 inch thick. Most carpenters use the standard formula described on page 11.

The brace table is located on the back of the tongue and is used for determining the precise lengths of braces needed for shelves or perhaps under the eaves of a house. It consists of 14 different calculations that look like this:

$$\frac{24}{24}\ 33^{94}\ \frac{27}{27}\ 38^{18}\ \frac{30}{30}\ 42^{43}$$

The two coupled figures, such as $\frac{24}{24}$, represent the length of the vertical and horizontal supports in inches. The next number, $33^{94}$, represents the length of the brace required—in this case, 33 and $^{94}/_{100}$ inches, or for practical purposes, $33^{9}/_{10}$ inches.

Keep your square clean and free of rust or its tables and calibrations will become difficult to read.

Cutting rafters so they fit together precisely is a critical part of framing a house. If they do not fit properly, the roof will be weak and out of square, which will result in additional problems when you put on the roofing material. With the aid of the framing square, all the necessary angled cuts and lengths can be accurately calculated in advance. Then you can cut two rafters and put them in place to check your work. If you are correct, you can use one of the rafters as a pattern to mark all the others. The principles of roof framing are discussed on pages 96–98. Instructions for using the framing square to make the necessary calculations are included there.

### Carpenter's Belt

If you are going to do any extensive carpentry, such as building a deck or a cabin, you might as well invest in a good tool belt immediately. A good belt is made of leather and normally includes four nail pouches, a ring to hang your hammer from, smaller pockets for chisel, nail set, and pencil, a pouch for the tape measure, and a slot for your combination square. A carpenter's belt saves quite a bit of time, since you can carry your most important tools and hardware with you instead of bumbling around trying to remember where you left them.

### Chalkline

A chalkline consists of a piece of string wound inside a container of colored chalk. The chalk-coated string is stretched tight between two points and then pulled up and allowed to snap back. It quickly marks a long straight line and eliminates the need for extensive penciling with a straight board. Most chalklines can also double as a plumb bob.

### Nail Set

A nail set is like a tiny spike and is designed to drive a finishing nail below the surface of the wood so it can be hidden. It comes in several different sizes for different finishing nails. An intermediate-size nail set should be in any carpenter's tool belt. You will need it to put up doorjambs and trim.

**Other Useful Tools**

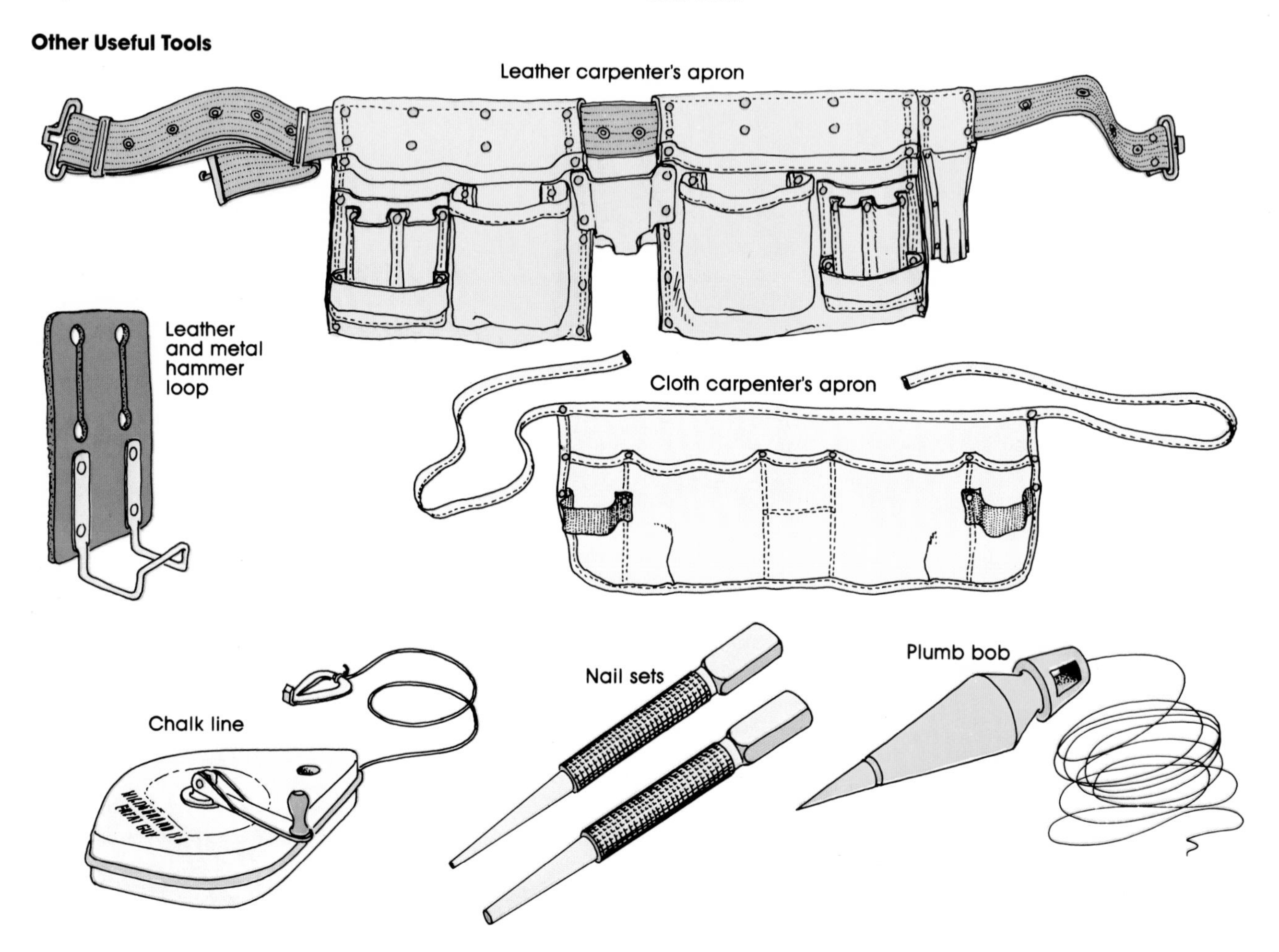

**Other Useful Tools**

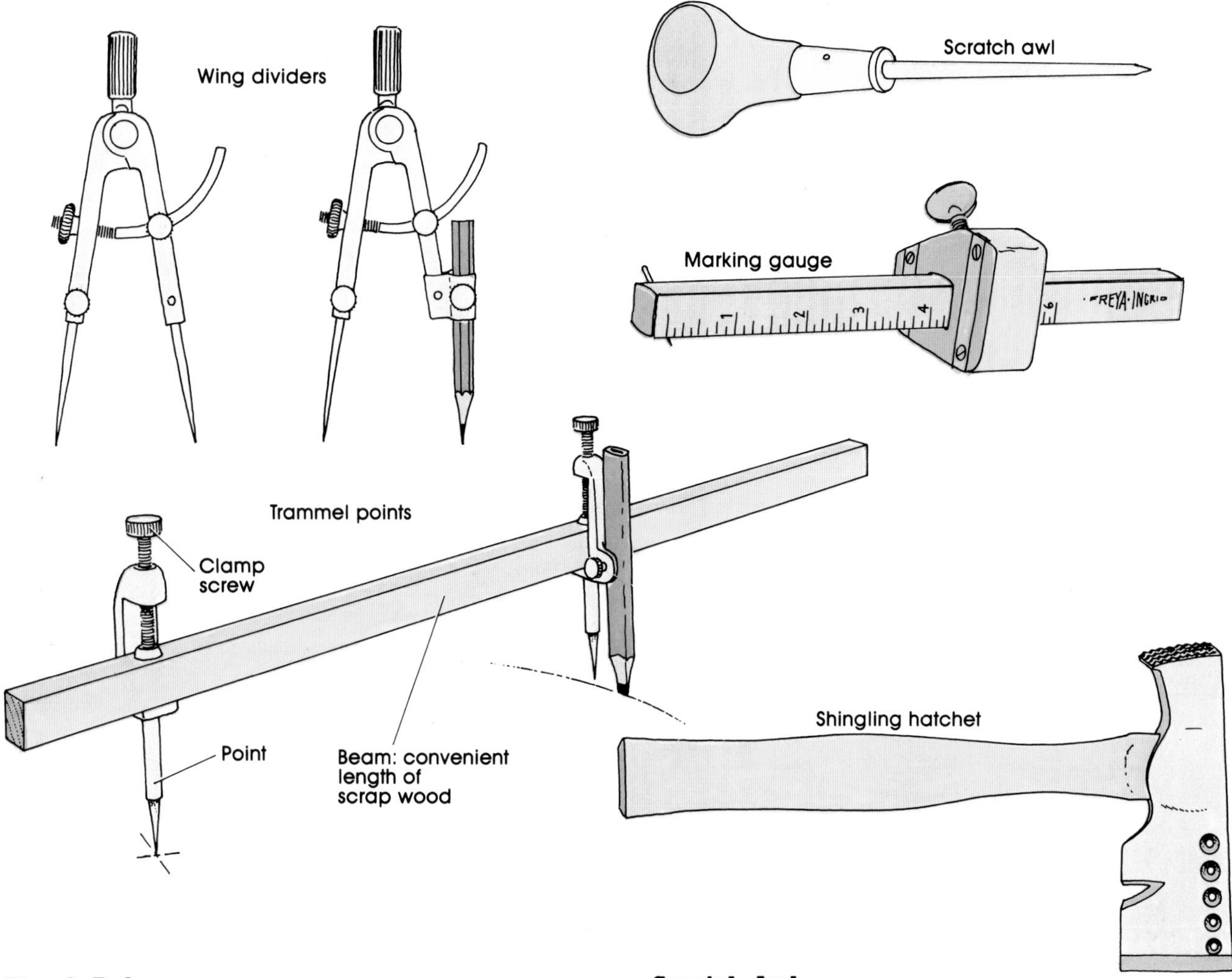


### Plumb Bob

The plumb bob is designed to hang from a length of string to give you a true vertical. With a true vertical and a square, you can find any horizontal. The plumb bob is widely used for locating the corners of a foundation. After the string is in place, the plumb bob is lowered next to the string to mark the corner posts (see page 80).

### Wing Dividers

Wing dividers are generally used to copy irregular lines. While they come in two styles—one with two points and one with a point and a pencil—the latter is most useful. For example, if you need to fit a long board against an irregular rock chimney, you can hold the board beside the chimney and follow the ins and outs of the rocks with the point while the pencil copies that line on the board.

Wing dividers can also function as a compass for drawing circles.

### Trammel Points

This is a specialty tool used primarily for marking a circle beyond the scope of wing dividers. Both points are clamped to a stick or ruler, with one point holding at the center and the other point moved out to the desired diameter of the circle.

### Scratch Awl

The awl, like a long, sharp nail held in a handle, is useful for making starting holes for small screws, or using instead of a pencil to mark lumber by simply scratching the surface. It is a useful but not essential item.

### Marking Gauge

The marking gauge is designed to make a line parallel to the edge of a board. One side follows the board's edge while the point, set at the desired distance, marks the surface of the board. Carpenters often do this by setting the distance with their combination square and then, with a pencil held at the end of the blade, sliding the square down the board. One type of marking gauge has double points and is used to mark mortises.

### Shingling Hatchet

The shingling hatchet is the all-purpose tool for putting on a shingle or shake roof. The hammer end drives the nails, and the hatchet end splits shingles or shakes for a perfect fit. The holes in the blade for the measuring gauge are set at ½-inch intervals. A gauge screws through the hole and quickly marks the amount of exposure between the shakes or shingles.

# SHARPENING TOOLS

### Folding Knife
Most experienced carpenters carry a good folding knife in one pocket. It handles a myriad of little chores like cutting string, sharpening a pencil, or even digging out splinters from your hand. Get a good one, keep it sharp, and you will have a tool for life.

### Utility Knife
Utility knives come in a variety of styles, but the common denominator is that each holds a razor blade or something similar. The blade can be retracted when not in use to protect it and to prevent accidents. The utility knife is used to cut string, roofing felt, insulation, and so on.

### Bench Knife
This knife has a set blade and remains around the workbench. Typically, a bench knife has a long and fairly large handle for an easy grip, and a small, fine blade of high-grade steel that holds a sharp edge. It is used for carving or trimming wood.

### Sloyd Knife
The sloyd knife is similar to a bench knife but has a larger blade. Carpenters like to use it as an all-purpose utility knife, particularly for cutting wood, leather, or plastic.

### Wrecking Bar
It is almost a foregone conclusion that you are going to make mistakes when you take on a big project like a cabin. This means you will have to pry apart some boards you have nailed together. A wrecking bar makes the difference between cursing and sweating while you struggle with a claw hammer or pulling the boards apart with relative ease. Wrecking bars come in different styles and sizes, but most have one curved end for pulling nails and one flat end for prying. All have slots at both ends for pulling nails. This is an essential tool around any large carpentry project.

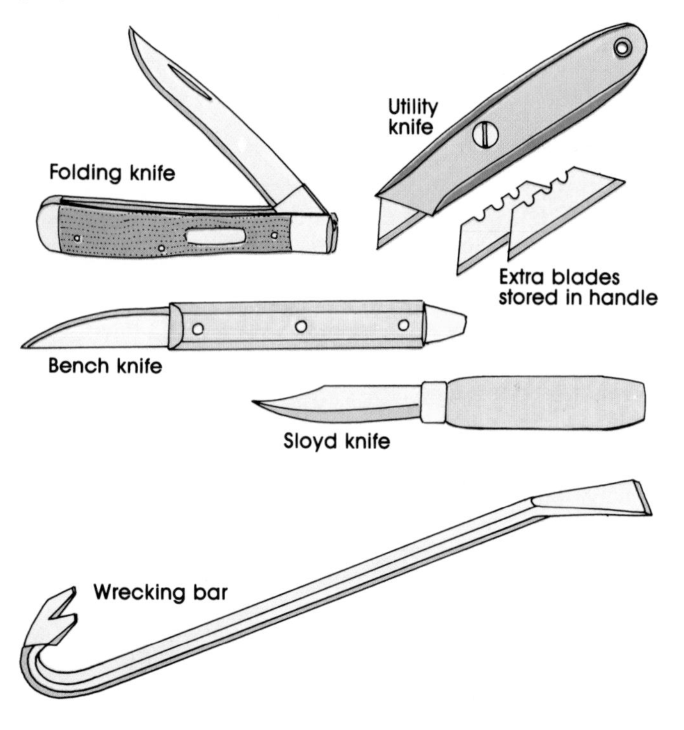


Dull or poorly maintained tools not only make your work harder, but are dangerous as well. It's the dull axe that ricochets off the wood, and the loose hammer head that becomes a lethal flying missile. A few minutes set aside every week to go over your tools and correct problems will prevent accidents and make your life easier.

Saws require very precise and careful work and are, therefore, best sharpened by the professional. Doing it wrong by yourself can ruin a good saw. A chisel or plane iron should be sharpened on an oilstone. A sharpening clamp, usually called a honing guide or sharpening jig, is helpful for holding the tools firmly at the proper angle. A properly sharpened chisel or plane iron should cut the hairs on your arm without pulling. Burrs on drill bits should be filed off, and bent bits thrown away. Always follow the existing bevel on drill bits.

## Oilstones
Fine, medium, and coarse oilstones (or whetstones) are available, and they can be used to sharpen just about any tool. An oilstone that is coarse on one side and medium on the other is often used around the workbench. The stone should not be less than 6 inches long and 2 inches wide, and twice this size is even better. A few drops of light oil placed on the stone while whetting a tool keep the pores of the stone from getting clogged.

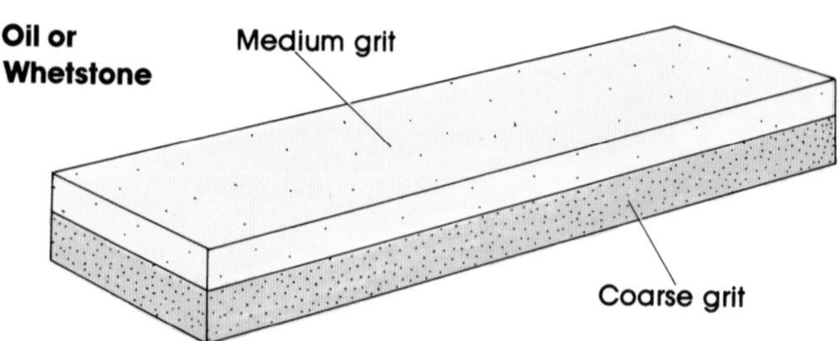


## Bench Grinder
The bench grinder is essential if you want to keep your tools sharp and clean. It's a fairly large investment for your shop but will give you years of good use. You might want to check out several different styles and prices in your hardware shop and then scout around flea markets and garage sales for a good deal.

Typically, a bench grinder has two grinding wheels, one medium coarse and one medium fine. A steel wire brush for cleaning and buffing is often interchanged with one wheel. A good bench grinder should also have adjustable plastic eye guards over each wheel and a small adjustable bench rest below each wheel.

Set the bench rest to the proper angle before you begin sharpening any tool. Then start the motor and move the tool evenly back and forth across the wheel. If you stop in one place, you will get an uneven grind.

Grinding heats metal and can remove its temper, or hardness. To prevent this, keep the tool cool by dipping it regularly in a nearby can of water.

Chisels and plane irons should be put on a wheel only when they are badly nicked or markedly ground down from repeated whetting. Don't try to sharpen them with small grinder wheels or you will cut a curve into the bevel of the chisel or plane iron.

# SAFETY

Many people preach about safety measures but don't practice them—until they get hurt. If you get hurt on the job, whether in your workshop or on a construction site, more often than not it is your own fault. Getting hurt on a job has two immediate effects: first, it's painful, and second, you may not be able to finish your project.

There are many pitfalls in carpentry. After all, you are working with sharp and heavy tools, often in awkward positions. The basic rule for practicing good safety is **stay alert.** That means watch what you are doing and watch what others around you are doing. It also means making sure that you are using your equipment and tools in the proper manner and that they are always well maintained. A few specific hazards and safety rules are described below.

## Circular Saws

Make sure the blade guard always moves freely and will snap back into position when you finish a cut. Never jam it up. When you have to raise it to do a certain type of cut, watch what you are doing.

Don't leave a power saw plugged in and unattended around a work site where a child might pick it up and turn it on.

Never make a cut with the power saw directly in front of your face. The blade can jam and kick the saw back into your teeth.

## Hammers

Be sure the head is always firmly fixed to the handle. If it is loose because the handle is excessively dry, soak it once in water to tighten it by swelling the wood. But don't keep repeating this process. The swelling and drying will soon make the situation worse. Drive another wedge in the top, and if that isn't enough, replace the handle.

If the handle is splintered and frayed just below the head, replace it before it breaks.

**Parts of a Bench Grinder**

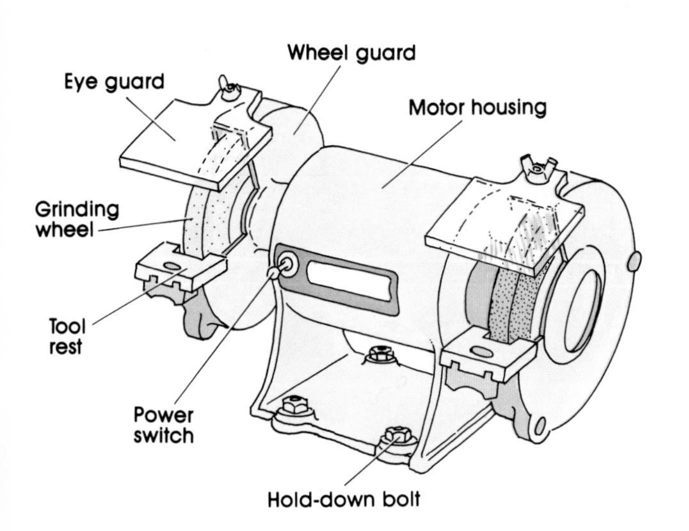


Keep the head clean and free of dirt or oil that will cause it to slip. If it is chipped or battered on the face, throw it away. Don't strike another hammer on the face with yours; because of the extreme hardness of the faces, you can send chips flying into your eyes.

## Falling Tools

If you are working on the roof or second story, keep all your tools in your belt. Never put your hammer on top of a stud wall or narrow board where a sudden jar might topple it onto the head of your partner below. The same goes for handsaws and all other tools.

When you are putting on a roof, make sure that all your equipment is secured and will not suddenly take a flying trip to the ground. That goes for you, too.

## Ladders

It's hard to convince anyone else that it wasn't your fault when you fell off a ladder. They will know you were careless. Make sure that the ladder is firmly anchored on the ground and will not slip away. If you are using a metal ladder on smooth concrete, glue some pieces of rubber or outdoor carpet on the feet to keep them from slipping.

Make sure that the ladder is planted evenly on the ground and stationed securely at the top. Do not go higher at the top of the ladder than your thighs. You must always be able to reach the ladder with one hand. If you must use a ladder in an awkward position, such as when climbing on or off a roof, have your partner steady it for you.

Keep the steps clean and free of grease or mud, and regularly check that all bolts or rivets are in place and tight. If a rivet breaks on a metal ladder, drill it out and replace it with a bolt, nut and washers.

## Axes and Hatchets

Keep these and all cutting tools sharp and clean. A sharp tool, whether it is a knife, chisel, or axe, cuts through the wood with less strain on the handle and will not be inclined to slip, as will a dull tool.

If you are using a double-bit axe around the job site, always lay it flat on the ground when it is not in use. Never leave it stuck in a stump or even leaning against something where someone could step or fall on the blade.

Always carry an axe or a maul high on the handle, next to the head. That way you can throw it away from you if you slip. Never carry it on your shoulder, where a slip may force the head into your neck or back.

## Electrical Cords

Don't allow extension cords to lie across wet areas. Breaking this rule can be shocking.

When you see a frayed or broken section in an electrical cord, wrap it immediately with electrical tape. If the cord is becoming frayed or broken where it enters your power tool, make a complete repair so no exposed wires show. Otherwise, sooner or later, you or someone else is going to grab the cord at that point and get a shock.

Remember the basics, and you shouldn't have problems: stay alert, use well-maintained tools, and use the right tool for the job.

# TECHNIQUES FOR SMALL CONSTRUCTION

A great way to practice basic carpentry techniques is to build small projects for use around your home and workshop. Shelving units, basic cabinets, a workbench, and sawhorse all offer plenty of opportunity to gain skills in measuring, cutting, and joining wood.

If you are an amateur builder, you should try some of the small projects described in this chapter before attempting the large-scale construction described in Chapter 4. Far from being make-work, these projects will provide you with useful items for your workshop while increasing your carpentry skills at the same time. The projects include a workbench, sawhorse, drawers, shelves, and cabinets. The instructions for building them are of a general nature and can be easily adapted to your own specific requirements. If you can plan and execute small projects like these successfully, you will be ready to tackle the cabin or playhouse described in Chapter 4.

## Planning the Small Project

Before you start any project, you need to draw up plans. Well-prepared plans not only show you how to put things together, but anticipate construction problems as well. Instead of trying to figure out each step on the work site, which is very time-consuming and inefficient, you should plan everything out on paper before you start, no matter how simple the project may seem.

For example, the first step in building shelves is to determine just where you want to place them. Will they be easy to reach? Will they be in the way of anything else? What is the most practical way to support them?

The next step is to make a series of careful measurements at the spot where you want to put them. Decide how high, how wide, and how deep you want the shelves to be. Make sure there are no obstructions, such as doors or windows.

Next, consider the type of wood you want to use. Are the shelves just for the workshop or will they be used in a more formal setting? The more formal, the better the quality of the wood and finish should be. If you plan to build shelves that are deeper than the widest standard board available (11¼ inches), you will have to put two boards together or cut them from plywood or particleboard. Once again, if you use plywood you will need to determine the quality you want (see page 9 for details on plywood qualities).

When you have decided on your measurements and wood qualities, make a sketch of the shelves and support system. Make the sketch to scale, using graph paper. Each square on the graph paper should represent a certain number of inches or feet of your project. When you have completed your drawing, double-check your measurements. Accurate measuring is crucial. If you cut a board

◀ Sanding gives a more finished appearance on small projects. See pages 56–57 for sanding techniques.

Accurate measuring is crucial to any project. This is especially true with drawer construction. See pages 70–71.

**Plans for Bookcase**

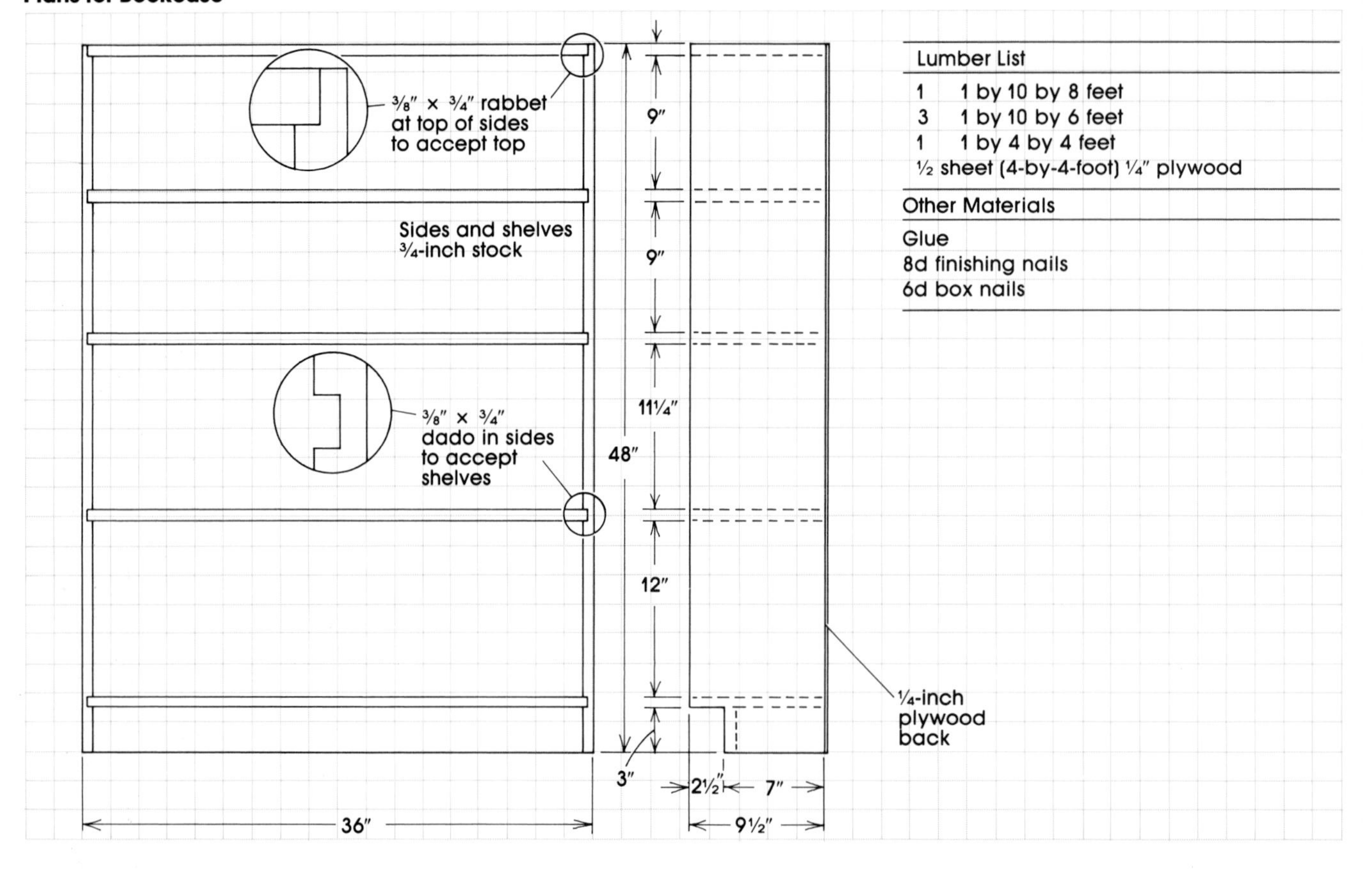


too long, you can always trim it down, but a board cut too short is wasted. Measure carefully.

Another thing to think about as you draw up your final plans is how to get the most out of the available wood. This means making the fewest possible cuts with the least possible waste. As an obvious example, you should avoid buying 12-foot boards for cutting pieces that will be 9 feet 6 inches long. Instead, buy 10-foot boards so you will waste only 6 inches. Good planning can be particularly economical when you cut up sheets of plywood for cabinets and drawers. Plan your cuts on graph paper so as little wood as possible will be wasted.

In cabinetwork, cut all pieces first and check that everything fits together smoothly before trying to assemble it. That first dry run allows you to correct any mistakes.

All these steps are essential to good planning. Detailed worksheets, complete with a list of all the materials you need, will allow you to proceed at the most efficient speed with the least number of problems. Without plans, you may make mistakes that will mean starting the project over.

Finally, a traditional word of advice for those who try to take too many shortcuts:

*A lazy person works twice as hard.*

## Basic Boxes

The knowledge required to build a simple box is fundamental to both basic carpentry and basic cabinetmaking. Even if you never build a house, most carpentry projects require a workshop complete with workbench, shelves, and drawers for storage. The basic box is the first step in building all of these items.

The basic box consists of four boards of equal length nailed and glued together. The variations described here may seem very elementary, but remember that a box with a door hinged to its front becomes a cabinet when it is hung on the wall or a toolbox when it is placed on the floor. Making boxes is one of the most practical ways to gain carpentry skill. For practice, you can build a variety of boxes—tall, short, fat, skinny—and stack them for an unusual bookcase, stereo cabinet, or pantry. (See page 38 of Ortho's book *Wood Projects for the Home* for an example of this idea.)

## The Butt Joint Box

The butt joint box is the easiest type of box to construct, although structurally the weakest. However, you can greatly increase the strength of the joints in any box by gluing them and by using screws instead of nails.

Decide what you will use the box for before you make it, since this will determine the size and length of the boards you need. If you want to store small jars of screws and bolts, 1 by 4s will suffice; if you need a spot to keep paperback books, use 1 by 6s. The larger the box, the more likely it is that you will need something to keep the box square, such as shelves or backing, both of which are discussed below.

After determining the size and length of the boards you need, cut the top and bottom pieces to equal length and the two side pieces to equal length. To assemble the box, lay it on a table or workbench and check that the pieces fit together. Spread a thin layer of white glue on the boards where they meet and then nail them together with finishing nails. Put the box together one piece at a time, and keep it as square as possible as you work. Once the box is nailed and glued together, place a square over its corners to check that it is square. If it isn't, push the corners slightly to square it before the glue begins to set.

Let the glue dry, then set the finishing nails below the wood surface with a nail set. Fill the holes with wood putty, and paint or stain your box. It is somewhat easier to paint or stain before assembly, but this is not practical when you have nail or screw holes to fill in and touch up.

Large butt joint boxes, such as for bookshelves, will probably not remain square. This is particularly true if they are made from 1-by stock (stock cut to a nominal thickness of 1 inch) rather than 2-by stock. One way to increase the rigidity of a box is to cut backing from ¼-inch plywood or ⅛-inch hardboard and glue and nail it to the back. Another way is to place one or more shelves within the box.

A simple and effective shelf can be made by cutting a board to fit snugly inside the box. Measure down both sides of your box so the shelf will be level. Square the box again, and keep it in place by pushing it against the wall at the back of the workbench while you glue and screw the shelf into position. Alternatively, you can glue and nail pieces of quarter-round molding to the sides of the box as shelf supports.

## More Elaborate Boxes

To make a box that is stronger and more attractive, more complex joints are required. For example, the butt joint can be replaced by the miter joint with only a little more skill. And instead of nails or screws, you can use dowels along with glue.

Cutting a miter joint does not take fancy tools, just careful work. You can find the 45 degree angles with your combination or try square, and you can make the cuts with either a handsaw or a power saw. (See page 22 for details on cutting a miter joint.)

If you want to give the box a truly finished appearance, you can use dado joints to attach the shelves. First, mark the sides where you want the shelves and cut the dadoes. Place white glue on the butt ends of the shelves and in the dadoes, then slip the shelves in place. Wipe off the excess glue, turn the box on one side, and weight the other side. This will press the shelves firmly into the dadoes until the glue dries. (See page 24 for details on cutting a dado joint.)

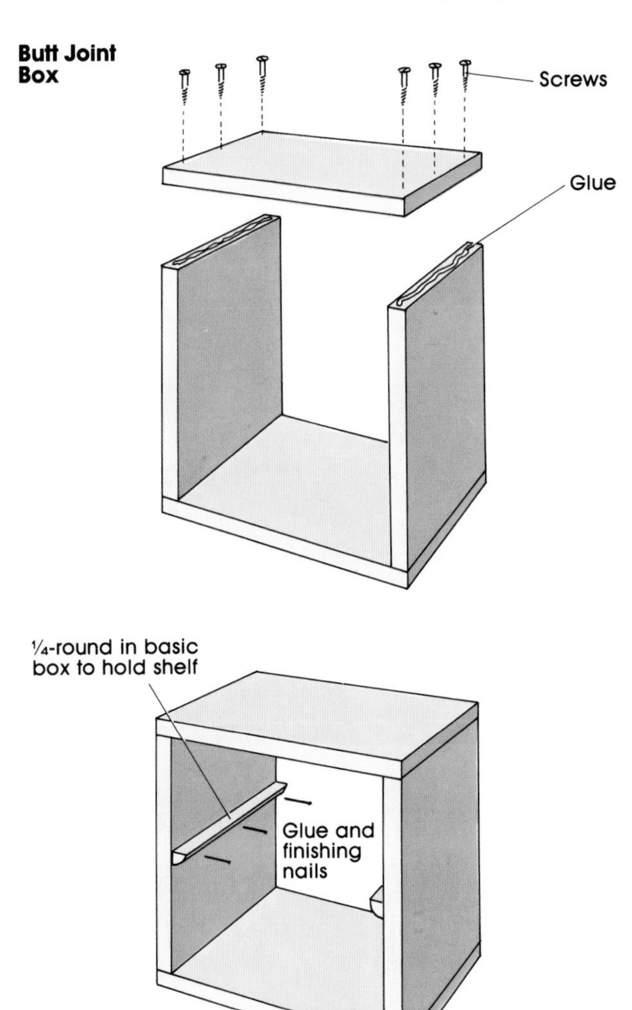


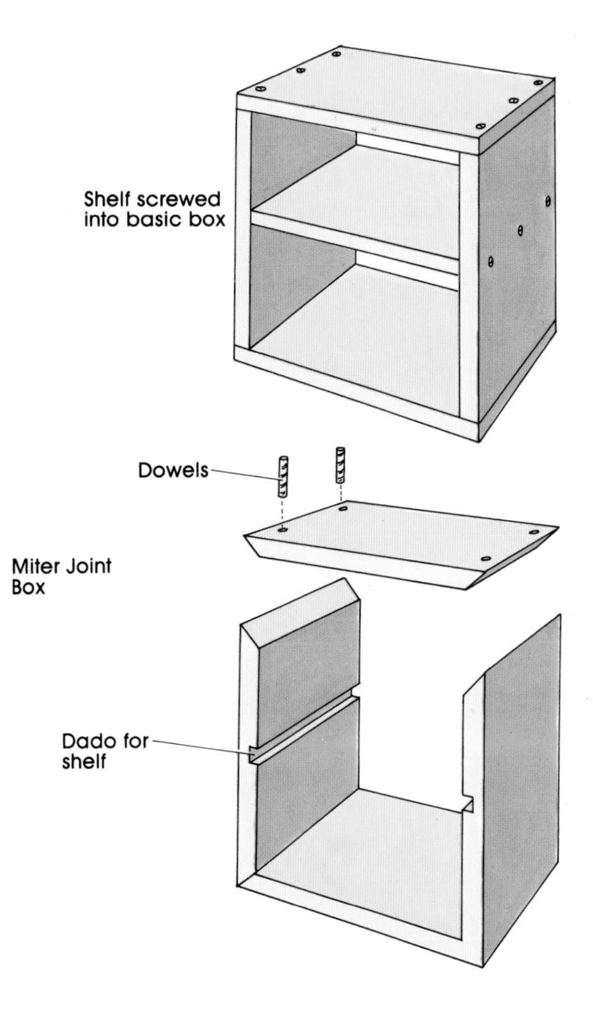

# WORKBENCH

There are as many varieties of workbenches as there are woodworkers. Even if you copy the design covered here, you will probably add some refinements to suit your personal needs—and that's the way it should be. Use the basic techniques shown here to design your own work area. Take into consideration how much room you have, what types of tools you have, how strong a bench you need, and how much time you plan to spend there.

## Fixed Workbench

The work surface of the fixed workbench is considerably higher than that of a portable workbench. The fixed bench is used most often as a work and repair area, and it needs to be higher to prevent excess stooping on your part. A good rule of thumb is to make it as high as the top of your belt—maybe an inch or two above that, but not lower. And don't make it too wide or you will have trouble reaching tools at the back. Comfortable widths range from 26 to 30 inches, and 32 inches is about the limit. The length of your bench is largely determined by how much room you have, but an 8-foot-long bench is large enough for most projects.

In addition to a good solid work surface, the fixed bench should provide ample storage area. A good combination is storage shelves on one side and drawers on the other end, with an open storage area in the middle.

The plans provided here are for a basic workbench that is 8 feet long, 27½ inches wide (five 2 by 6s), and 40 inches high. If you plan to put in drawers, make them before you make the rest of the bench (see pages 70–71 for details). It's impossible to alter the drawers once they are made, but you can adjust the drawer space to accommodate them.

Here are the basic steps in construction of the fixed workbench:

**1.** Cut eight legs from 4 by 4s. Each leg should be 38½ inches long.

**2.** Cut eight crosspieces from 2 by 4s. Each crosspiece should be 25 inches long.

**3.** Construct the two outer leg systems. Cut one 1½-inch-deep dado and one 1½-inch-deep rabbet in each of four of the 4-by-4 legs (see pages 24–25 for details on cutting dadoes and rabbets). These cuts should accept the 2-by-4 crosspieces, as illustrated. The top crosspieces should be flush with the tops of the two legs, and the bottom crosspieces should be 10 inches from the ground. Glue and screw the outer leg systems together.

**4.** Construct the two inner leg systems. Position the crosspieces in the same manner described in step 3, but do not set the 2 by 4s into the 4-by-4 legs. The bottom crosspieces on the inner legs become shelf supports.

**5.** Stand the two outer leg systems up with the crosspieces to the outside. The leg systems are large enough to steady themselves. Glue and screw the five 2-by-6 boards that comprise the bench top into position, allowing a 12-inch overhang on each end. Put on the back board first, flush with the back edge. Screw two lag screws through the ends of each board into the top crosspieces. Countersink all lag screws. Make sure each board fits snugly against the next so there will be no spaces through which to lose small screws and nails. When you reach the front, you should have about a 2½-inch overhang.

**Fixed Workbench**

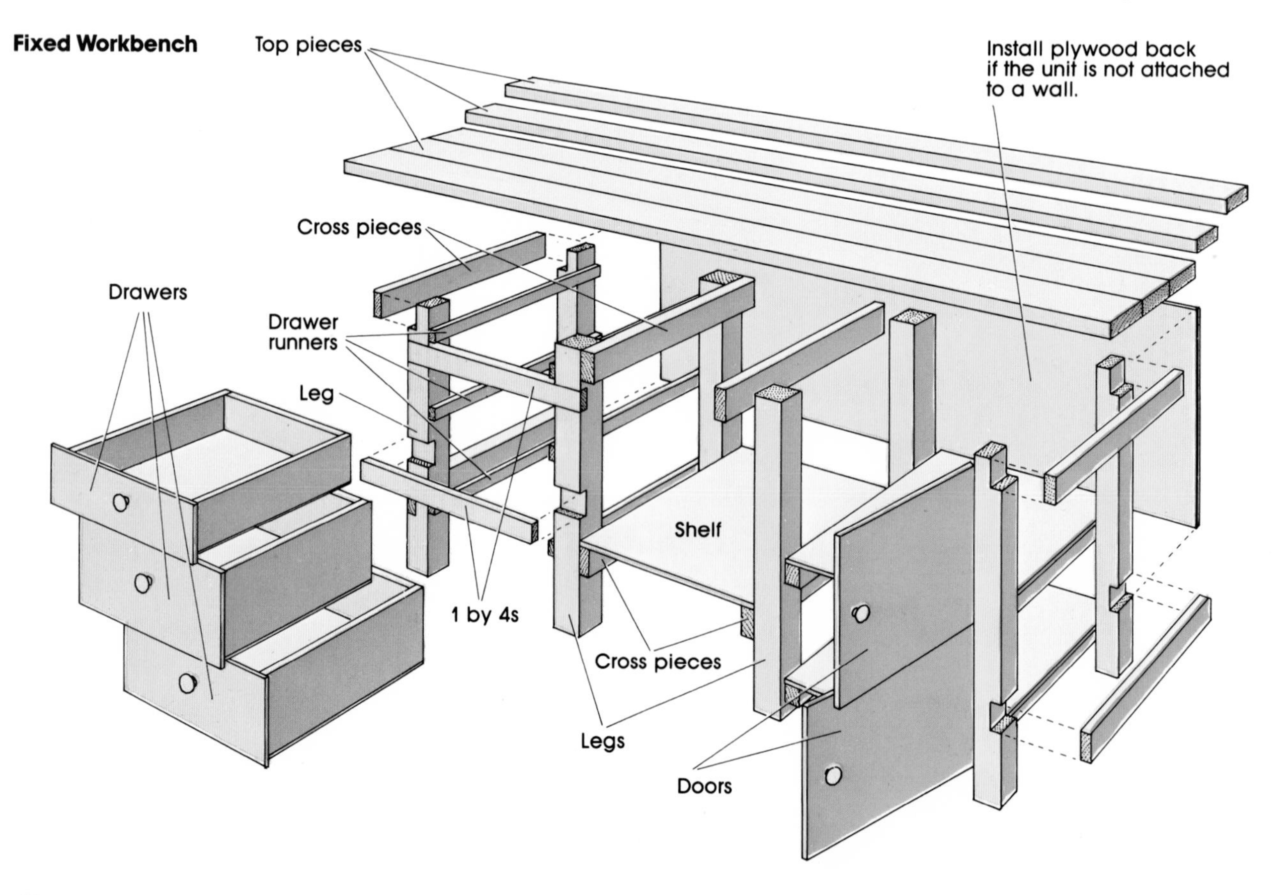

### Plans For Fixed Workbench

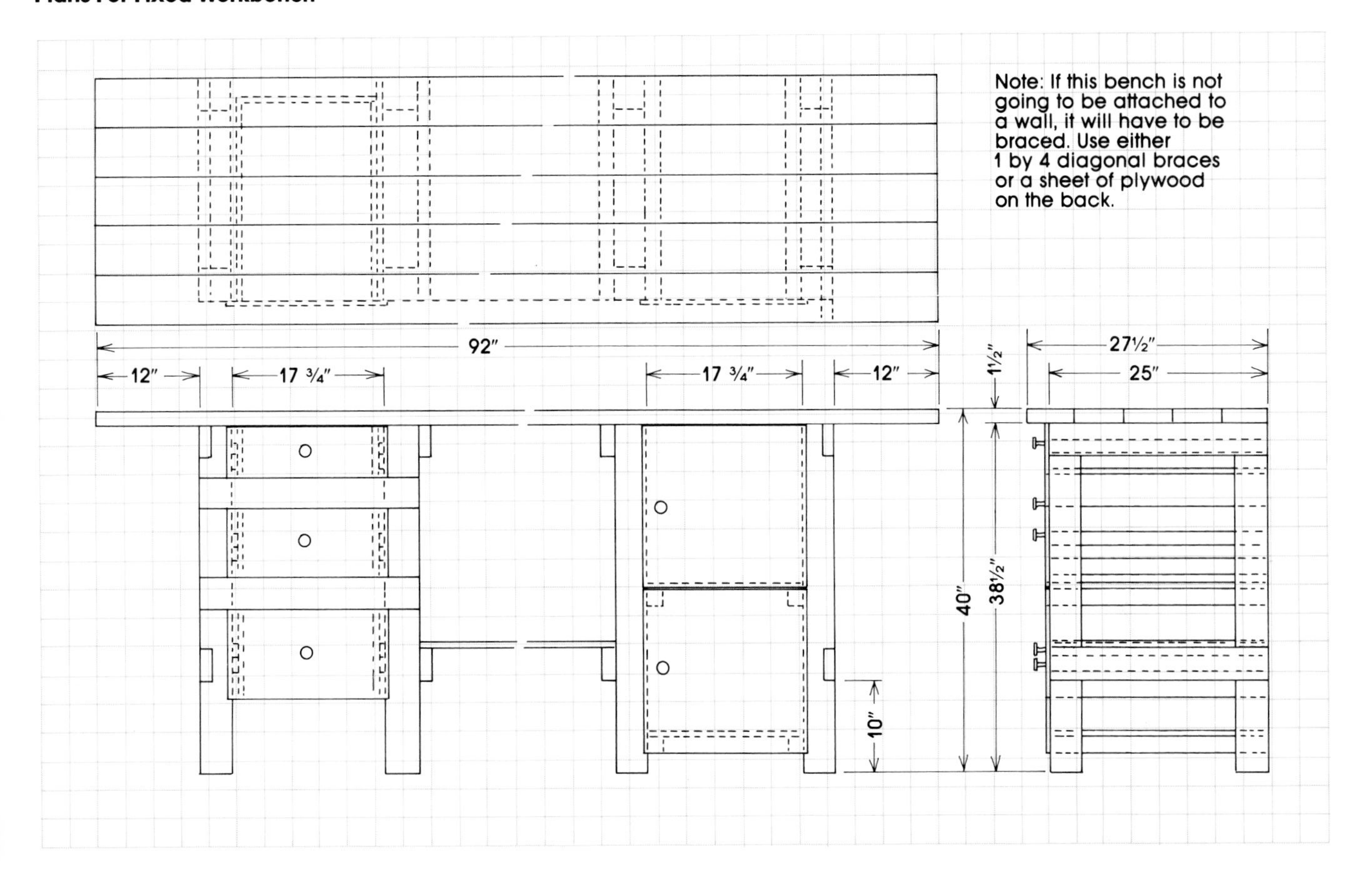


**6.** Position the inner leg systems. The drawers for this particular bench are 16 inches wide, but they ride on 1-by-2 runners nailed to the inner sides of the legs. These runners project out ¾ inch on each side, so the space between the legs is 17¾ inches (16-inch drawer width plus 1½ inches for two runners plus ¼-inch clearance (⅛ inch on each side)).

Carefully space the inner leg systems to provide an exact 17¾-inch measurement from the outer legs in both the front and the back.

**7.** Drive two or three nails through the top boards into the top crosspieces to hold them in place temporarily, and screw lag screws through the top into the crosspieces. Countersink all lag screws.

Use a steel square to check that all the legs are plumb and square. Double-check that the distance between the outer and inner leg systems is the same at the top and the bottom.

**8.** Make runners for three drawers. You can make the drawers any size you wish, but in this example the top drawer is made from 1 by 6s, the middle drawer is made from 1 by 8s, and the bottom drawer from 1 by 10s to provide varying depths. Each drawer is 16 inches wide and 23 inches long. The drawer fronts are overlapped with ½-inch plywood that extends out 1 inch in all directions to cover the gap left by the drawer runners. A 1-by-2 strip is glued and nailed to each side of each drawer. The bottom of each strip should be exactly in the center of each side. (See pages 70–71 for details on drawer construction.)

In order to position the support runners on the legs, hold the top drawer in the opening, allowing a ¼-inch clearance at the top. Mark on the two front legs where the bottoms of the drawer runners are to be positioned. Use a straight board and a level to carry these marks to the two rear legs. Glue and nail the 1-by-2 runners into position so that their top edges are flush with these lines. Wax the runners and insert the drawer.

Repeat this process for the other two drawers, leaving about ¼-inch clearance between each.

**9.** On the other end of the bench, make one or two shelf supports from 2 by 2s (wider than drawer runners for more support) in the same manner as the drawer runners. Glue and nail ½-inch pieces of plywood into position as shelves.

**10.** Make the cabinet door from ½-inch plywood. Cut it so it overlaps the opening by 1 inch on each side. The bottom of the cabinet door should also extend 1 inch below the bottom shelf. Use overlay hinges for ½-inch plywood. Mark 3 inches from the top and bottom of the door, center the hinges on these lines, and install them on the door. Center the door over the opening and screw the hinges to the outer legs. Use a magnetic catch to keep it closed.

**11.** Make a shelf across the center opening by gluing and nailing a piece of plywood or some boards to the bottom crosspieces.

**12.** Cover the ends and the back of the bench with plywood to keep dust and dirt from getting into the drawers and shelves. Paint or stain everything except the top, which should be left natural.

# DRAWERS

Drawers are stumbling blocks for many carpenters. It is very difficult to make a finely crafted drawer with dovetail joints all around. So start out with the easiest forms, and don't get nervous. After all, a drawer is just a flat box.

A drawer should not be too long or you'll never see what's kept at the back; it shouldn't be too deep or everything will be buried. As a rule of thumb, build drawers no longer than 30 inches and no deeper than 12 inches.

Just keep these three simple rules in mind and you won't go wrong: (1) the drawer must be accurately measured and cut, (2) it must slide easily, and (3) it must be strong. Beginners sometimes have difficulty with the measuring and cutting. Remember that the drawer must fit into a narrow space, so work carefully.

Drawers are usually constructed in the following sequence:

**1.** Determine the exact size of each drawer. If the drawers are different depths, remember that the shallowest one should be at the top.

**2.** When you consider the size of the drawer opening, plan ahead for the additional space that might be needed for the sliding mechanism.

**3.** Measure and cut the sides to equal length, and the front and back to equal length. The front and back normally fit between the sides. This helps prevent sliding objects in the drawer from knocking them out.

**4.** Make a face for the drawer that is large enough to cover all the gaps around the drawer opening. This is called overlapping the front of the drawer.

## Drawer Fronts

Drawers fit into openings in one of two basic ways: flush or overlapped. A flush fit requires very careful work to prevent large gaps around the drawer or too tight a fit. An overlapped drawer front, which has a face that is larger all around than the drawer and the opening, hides any gaps created by the sliding mechanism or by defects in your workmanship. It also provides a stop for the drawer.

One way to overlap a drawer front is to make it from a piece of wood that is larger than the drawer, as shown in the illustration. An easier way that is just as attractive is called the paste-a-face method. The larger face, usually made of plywood, is attached after the drawer is finished, using glue and screws that are driven from the inside of the drawer. Just make sure the screws don't go all the way through the facing.

## Drawer Construction

Aside from dovetail joints, which are beyond the scope of this book, there are three basic ways of joining drawer components together: the butt joint, the dado joint, and the rabbet joint.

### Butt Joint Drawer Construction

This method is the easiest and is therefore the best to start with. Technically, the butt joint is the weakest of the three types, but it is perfectly serviceable for workshops or basic cabinet drawers. Although the drawer described here is intended for use in the fixed workbench described on pages 68–69, you can adapt the basic plan to your own needs.

All the drawers for the fixed workbench are made from 1-by stock, regardless of size. Construction proceeds as follows:

**1.** Cut two 23-inch pieces for the sides and two 14½-inch pieces for the front and back, which fit between the sides.

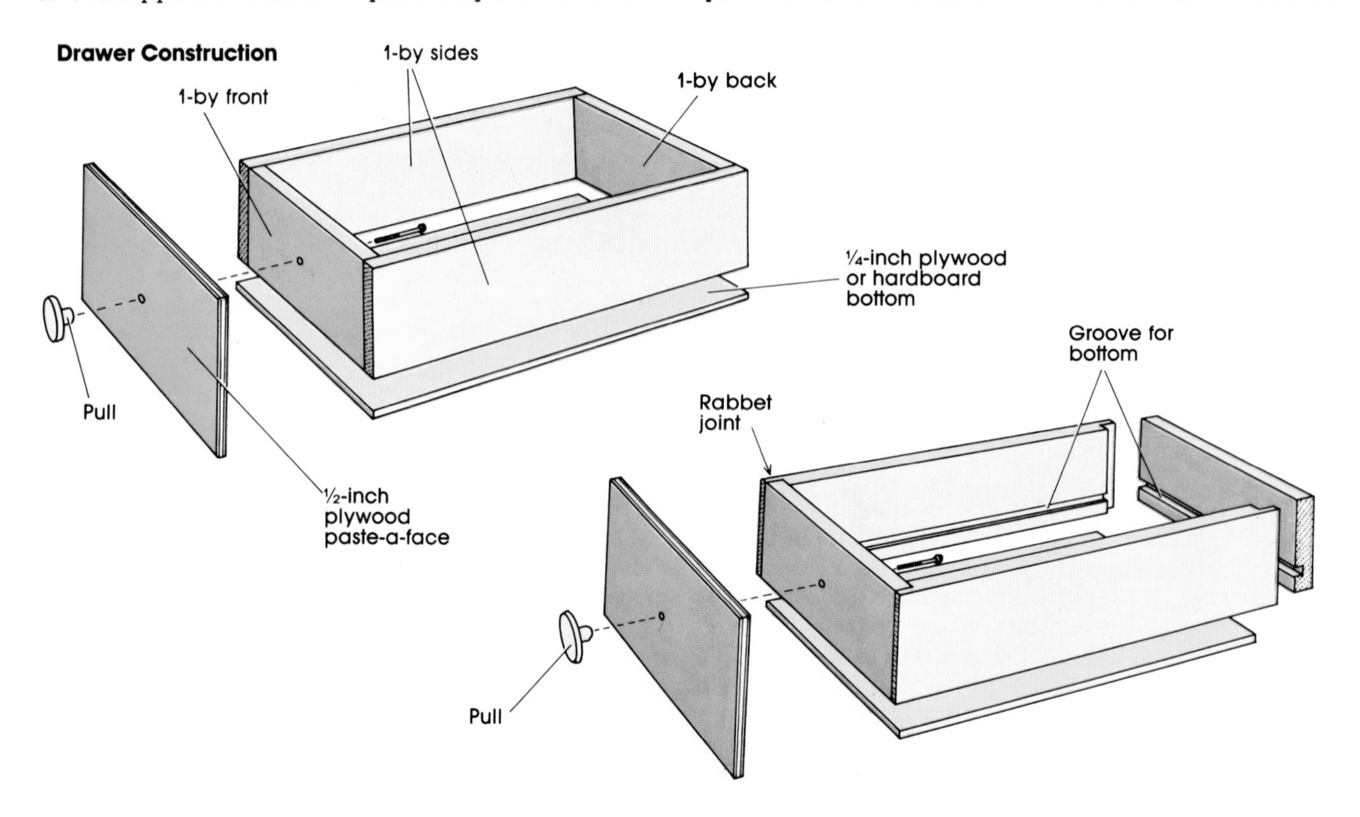

**Drawer Guides**

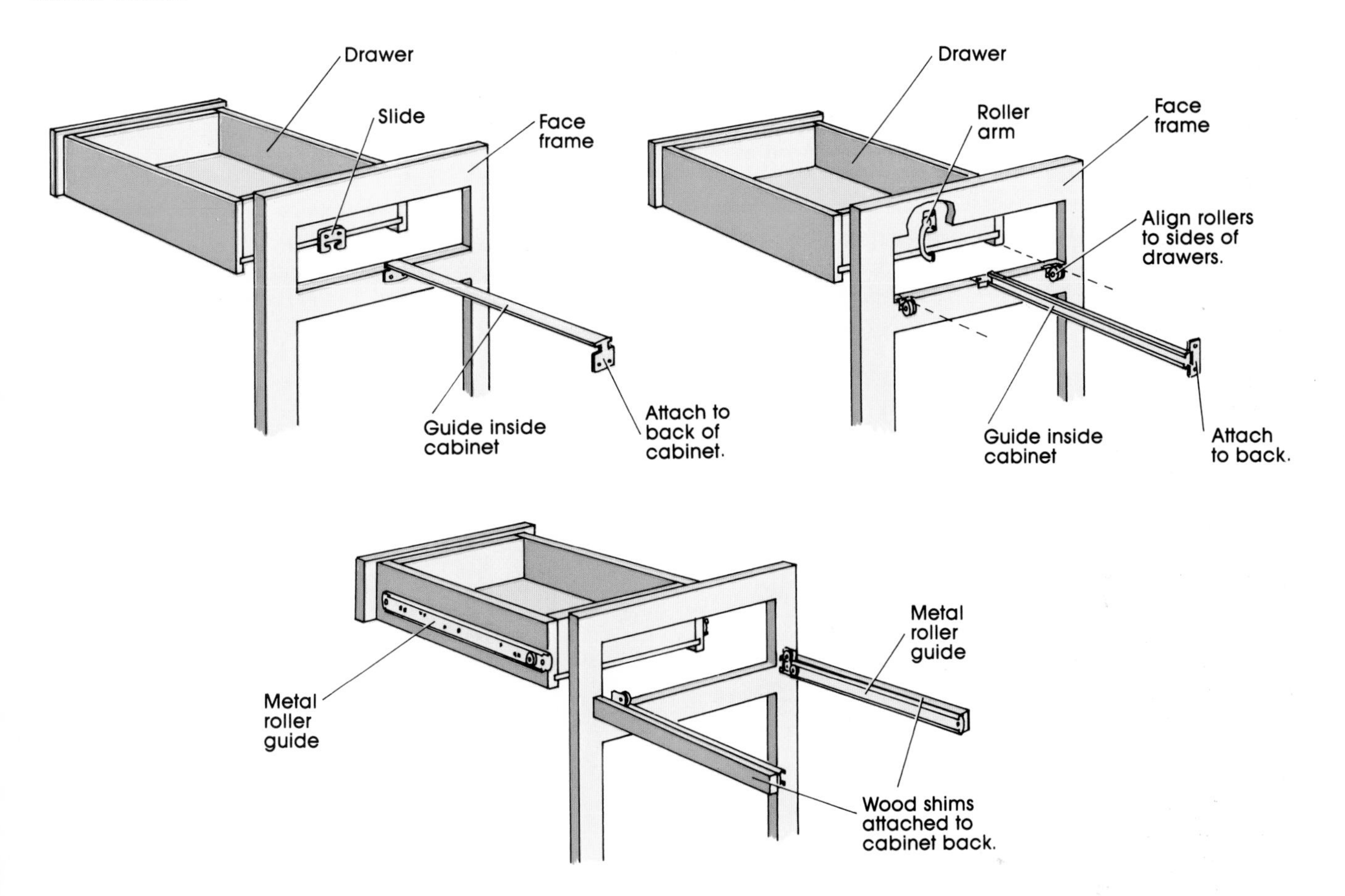


When assembled, the drawer is 23 inches long and 16 inches wide.

**2.** Glue and nail the front and back pieces between the two sides, working on a flat surface to keep all edges even.

**3.** For the drawer bottom, cut a piece of ⅛-inch hardboard and glue and nail it in place.

**4.** For the runners, glue and nail a strip of 1 by 2 on each side. The bottom of each strip should be exactly in the center of each side.

**5.** For the overlapping face of the drawer, cut a piece of ½-inch plywood 1 inch larger all around than the front of the drawer. Attach it to the front of the drawer with glue and four screws driven from the inside.

**6.** Put on the drawer pull of your choice.

## Dado and Rabbet Drawer Construction

A drawer assembled with dado and rabbet joints, which lock the pieces together, is very strong. However, it also involves much more precise work than simple butt joint construction. Construction proceeds as follows:

**1.** Cut the sides to length. Use a saw or router to rabbet the ends to accept the front and back pieces.

**2.** Cut the front and back pieces and try a dry run assembly to see that all pieces fit.

**3.** Cut grooves ½ inch up from the bottom on all four pieces. The grooves should be wide enough to accept a ¼-inch hardboard bottom and should be one-third the depth of the stock you are using for the drawers.

**4.** Cut the hardboard drawer bottom, allowing for the depth of the grooves all around. Try another dry run assembly.

**5.** Place glue on all joints and grooves *except* where the bottom fits into the grooves on the side pieces. This part is allowed to "float" for wood expansion and contraction. After the pieces are glued and assembled, put in finishing nails while the glue is still wet. Then let the glue dry thoroughly.

**6.** Attach the runners and the overlapping face in the same manner described for the butt joint drawer.

**7.** Attach the drawer pull of your choice.

## Drawer Guides

Drawer guides can be purchased in most hardware stores and attached to the drawer and its opening. A word of caution: the more complicated the assembly, the more difficult it is to get the drawer in straight and true. Your selection should be guided by your own woodworking experience.

Each assembly comes with its own instructions, but buy the guides you want before you make the drawers. Some of these assemblies require up to ½-inch clearances around the drawers, so read the instructions carefully before putting your drawers and cabinet together.

# WALL CABINET

Wall cabinets for storing tools, supplies, and other equipment can save a lot of space. The workshop floor is left uncluttered, and most of the stored items are at eye level and within easy reach. Closed cabinets also keep the items free of the dirt and sawdust that collect on open shelves, and can be locked if necessary. Best of all, a cabinet is not difficult to construct; it's just a box with a front cover.

## Basic Cabinet

One or more simple but serviceable cabinets can be cut from a sheet of ½-inch plywood. The first step is to make the basic box. Cut the sides, top, and bottom to your dimensions, install the shelf supports, then glue and screw the box together. Use ½-inch plywood for backing, to provide firm support for the cabinet.

While the basic box with a door attached is perfectly serviceable for the shop, the cabinet can be given a professional touch and placed anywhere in the house by making a facing for it from 1 by 2 pine or Douglas fir. This facing not only hides the rough edges of the plywood but provides solid material for driving screws.

To make the facing, cut the two side pieces (stiles) to run the full height of the box opening. The crosspieces at top and bottom (rails) are butted between the stiles so that

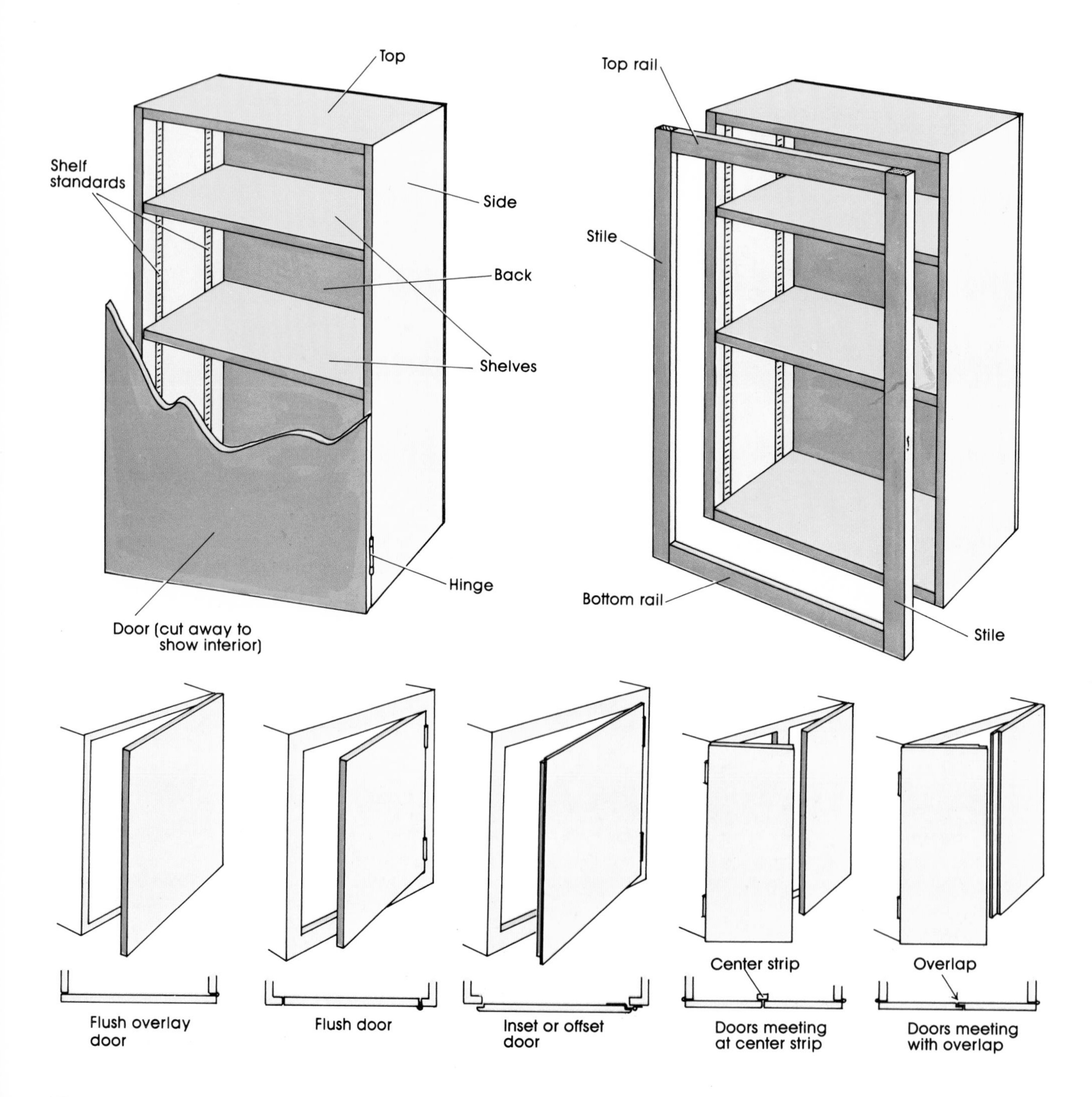

end grain is not exposed. Using glue and finishing nails, attach the stiles first and then the rails.

## Types of Hinged Doors

Doors to cabinets can be single or double; overlapped, flush, or lipped; made from any type of wood (although most are made from plywood); and can be hinged and latched in a variety of ways. The simplest type is a single overlap made with plywood, hinged with an overlap (overlay) hinge, and held closed with a magnetic catch or self-closing hinges.

The basic types of hinged cabinet doors are described below.

**Overlap door.** The overlap door is the easiest door to make. It is simply a matter of cutting accurately so that the door covers the opening. If the cabinet consists of a basic box without stiles and rails, the overlap door covers the entire opening. With stiles and rails, it should overlap on all sides from ½ to 1 inch. (See page 69 for details on the installation of an overlap door on a workbench.)

**Flush door.** The flush door is one of the most difficult to make. Because it fits inside the cabinet opening, it must be cut and hung precisely so it will neither bind nor have large gaps between its edges and the cabinet. Measure the opening very precisely, and if it is not square, adjust the dimensions of the door to fit.

**Lipped door.** The lipped door gives the cabinet a professional finish. It is normally made from ¾-inch stock and is cut so that it is ⅝-inch larger than the opening. The lip, which fits both inside and over the stiles and rails, is normally made by routing out a ⅜-inch rabbet around the inside edges. When the door is installed, these dimensions will leave a 1⁄16-inch clearance so the door will close without binding.

A lipped door can also be made by gluing two panels of ⅜-inch plywood together, the inside piece ⅜ inch smaller on all sides than the outer one.

**Double doors.** One door will suffice for a basic cabinet that is not more than 2 feet wide. A door wider than 2 feet will place undue strain on the hinges. If your cabinet is wider than 2 feet or wider than it is high, it is best to use two doors. The use of a center divider with double doors provides additional rigidity for the cabinet and a place for the doors to meet.

If you don't use a center divider, cut the doors so that they overlap by ½ inch and cut ⅜-inch rabbets on opposite sides where they meet for a smooth fit.

### Installing Hinges

The basic steps for installing most hinges are as follows:

**1.** Measure 3 inches from the top and bottom of the door. Place the hinges on the door, centered over the marks. Keep the hinge joint tight against the edge of the door and use an awl to make starter holes for the screws.

**2.** Screw the hinges to the door. Make sure the hinge joint remains tight against the edge.

**3.** Place the cabinet door over the opening and make sure it is even all around. For flush doors, use paper shims (spacers) to raise the bottom of the door off the cabinet.

**4.** Use an awl to make starter holes where the hinges rest on the cabinet, and screw the hinges to the cabinet.

**Sawhorse**

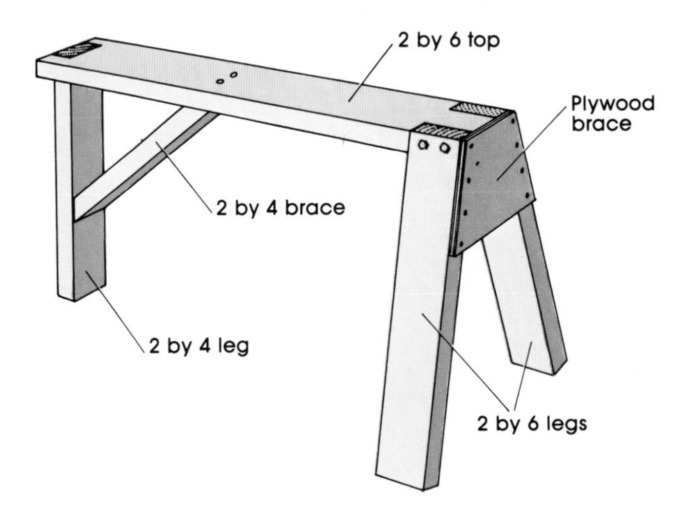


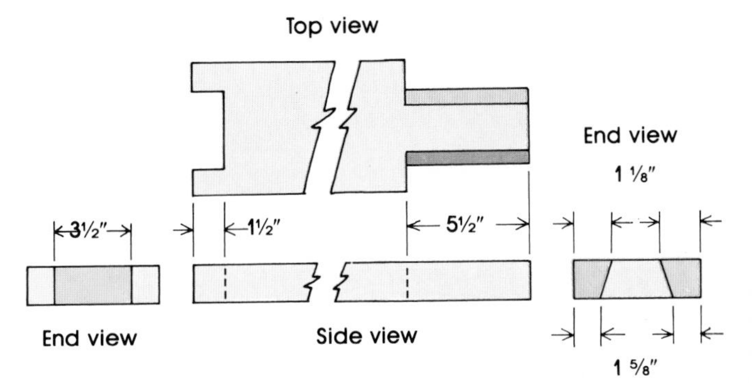


Working around the shop or on a construction site without at least two sawhorses makes your life considerably more difficult than it needs to be. Sawhorses are light and can be easily transported wherever you need them. Two of them placed near each other will balance any large pieces of plywood you may have to cut and will form a convenient scaffolding with a few boards laid across them.

A sturdy sawhorse that doubles as a tool storage unit and work center is described on page 99 of Ortho's book *Basic Home Repairs*.

The style described here works well on uneven ground, which is often a problem around construction sites. The basic steps in construction are as follows:

**1.** Cut a 4-foot piece of 2 by 6 for the top. Notch one end in the center to accept a 2-by-4 leg.

**2.** On the other end, cut two notches 5½ inches deep using the measurements shown in the illustration.

**3.** Cut two 31-inch pieces of 2 by 6 for the legs, each beveled to 30 degrees on one end. This will match the angle of the two notches for the legs.

**4.** Install the two legs with glue and lag screws. Don't screw them too tightly or you may cause a compression fracture.

**5.** Cut a 30-inch piece of 2 by 4 for the other leg and install it in its notch with two lag screws and glue.

**6.** Cut a 2-foot piece of 2 by 4 for the brace, with both ends beveled to 45 degrees. Glue and screw it in place.

**7.** Cut a brace for the other end from ½-inch plywood and nail it across the legs as shown.

# TECHNIQUES FOR LARGE CONSTRUCTION

From making a plan to putting on the roof, the construction of a building is nothing but a step-by-step process. Aided by the clearly stated techniques and detailed illustrations in this chapter, you can build your own playhouse, studio, or any project you have in mind.

A contractor friend once confided: "If people knew how easy it is to build a house, we would all be out of business." He was only half joking. Of course, not just anyone is going to go out and build his or her own house. It takes time, money, effort, and skill. But if you have the first two elements, and will make the effort, you can achieve the skill.

It's understandable if you feel somewhat overwhelmed when you first look at a house that's in the process of being framed. Where the contractor sees an orderly series of steps, you see only a maze of lumber standing on end. This chapter is designed to replace such confusion with a step-by-step understanding of how a house is built. However, the emphasis is on the *techniques* of house construction, not just on a particular set of plans. If you understand the techniques involved, you can build any kind of house you want.

As you read through this chapter, you will notice that it describes several alternative ways of approaching certain aspects of construction. This may seem confusing at first, but you should be aware of all the alternatives so you can choose the one that best fits your needs. Read through the entire chapter before you begin building to familiarize yourself with all the techniques.

The house constructed as a model for this chapter was designed as a playhouse. Its floor measures 8 by 12 feet, and the heights of the walls and doorway were scaled down to keep it in proportion. But by using the techniques described here, you can expand this house to 12 by 24 feet or larger for use as a guest house or vacation cabin.

Building a house is certainly not a piece of cake. It's hard work, and you must pay careful attention to detail, making sure that everything is square and plumb. If it's not, you will end up working twice as hard because nothing will join together properly and you will make numerous time consuming changes to make things fit. But if you make careful plans and follow them with accurate carpentry work, the whole project should be enjoyable for you and the friends or relatives enlisted to help you.

A word of warning: this book shows you how to frame and put the roof and siding on a house. You will feel great when that's accomplished; the house seems almost finished. In fact, you are less than half done. There's still wiring, plumbing, interior wall paneling, interior doors, ceilings, flooring, and countless other details to attend to. But at least you will have a dry and protected place to work in—so why not get started?

Construction steps in this chapter show you how to build this playhouse. The gable over the door is a nice variation.

**Whether you're cutting joists, studs, or rafters, the versatile circular saw is indispensable for large construction.**

# THE PLANNING STAGE

No matter what project you have in mind—building a playhouse, a cabin, a small addition to your house, or remodeling some portion of your house—you will need to know the techniques described so far in this book. Additionally, you need to be able to visualize all the steps that make up the construction process. The steps in this process are as large a part of carpentry technique as skill in using tools. If you take each step one at a time, follow good plans, and persevere in the face of frustration, you'll do a good job. The rest of this book describes and illustrates the techniques involved in the building process itself.

Before you start, you need plans or working drawings that show all stages of the construction process. There are several ways to approach this. You can design the project yourself, you can purchase ready-made plans for a particular type of structure, or you can have an architect draw up plans for you.

The first choice will only cost you your time. But if you want plans that are more professional than those you can draw yourself, there are dozens of companies across the country that will supply them at a moderate cost, anywhere from $5 to $50. Several popular magazines carry a few basic plans in each issue, along with information on where to obtain complete plans. Large bookstores usually have a good selection of books containing detailed construction plans for everything from room additions to barns and log cabins. And excellent plans for simple houses can be obtained by writing the Superintendent of Government Documents, Washington, D.C. Look over the widest possible variety of plans and then make your choice.

The last option—having an architect draw up the plans for you—is the most expensive. It can cost several hundred dollars.

### Prepackaged Homes

If you want to build your own vacation cabin or permanent home but are not sure that you are ready to frame a building yourself, you can look into the variety of precut homes on the market. These vary widely in style and price. Some come with all the pieces cut and marked for you to assemble; some come with whole sections already built, including the roof and floor trusses and the walls. Most of them require that you put down the foundation first and will send you the specifics for that with the plans.

Precut log cabins that go together like big Tinker Toys are also rapidly gaining in popularity.

## Reading Architectural Drawings

If you draw up your own plans, presumably you will be able to follow them. For a playhouse or simple cabin in the woods, they need not be too detailed. But if you are planning a more elaborate building or room addition and have the plans drawn up by an architect, you need to understand some of the basic steps in following the plans.

The plans consist of several pages of drawings and are not necessarily in any specific order. The first page may give you an artist's drawing of what the completed building looks like. The next page will probably be a bird's-eye view of the floor plan. It will include the overall dimensions of the building and door and window locations, including the size of the windows. Windows are sized with two sets of numbers, such as 4'0" × 3'6". The first number always refers to the width of the window, and the second to the height. At the hardware store, you would refer to this as a "four-oh, three-six window." The bird's-eye view also shows your plumbing and electrical fixtures. Overhead lights are usually represented by the symbol ⊠ , and wall outlets by the symbol ⊖ .

There may be smaller drawings on each sheet called details. These give specific information on certain aspects of the house, such as the layout of the roof, including the type of roofing material and the direction of water flow.

Other pages will show the front, rear, and side elevations of the building—that is, the walls. These pages will specify the size of the studs and the type of wall covering. More details will precisely define the foundation, including the depth of the footings, which varies according to local codes and the depth of the frost line.

Some of the last sheets will show construction details that tell you just how the architect envisions putting the building together. They seem intricate, but remember that it's all done one step at a time.

### The Lumber List

The lumber list is a complete breakdown of all the lumber, plywood, and nails required to frame the building. Some architects include it with their drawings, while others rely on the contractor to make the estimate. For a simple cabin or playhouse such as the one described in this book, you can make your own list. After you have finished all your detailed drawings on each stage of the construction, from foundation to roof, make a list of all the components needed. This list serves two purposes. First, you can just hand it to the lumberyard employees, and they will locate all pieces and deliver them to you. Usually there is no charge for this on such large orders. Second, with all your materials on hand, you don't have to waste hours running back and forth to the lumberyard for little pieces you forgot about.

If you have doubts about your lumber list, give your plans to an experienced employee at the lumberyard and the job will probably be done for you, since they want your business.

## Checking the Building Site

With your plans in hand, start walking around the building site. Here are some things to look for:

■ Is the ground firm and fairly level? If there are wet and spongy spots, you may have some type of seepage, either from a pipe or a natural spring. Not only will the ground be too soft but you will have constant moisture under your house.

The more level the ground, the easier it is for you to build. You may want to bring in a grader or bulldozer to level the building site. Foundations are often placed on steep slopes, but that is beyond the scope of this book. If you must build on a slope, consider having professionals put in your foundation.

■ Which direction do you want the house to face? A view is one consideration, but so is a southern exposure. Most

**Drawing Up Plans**

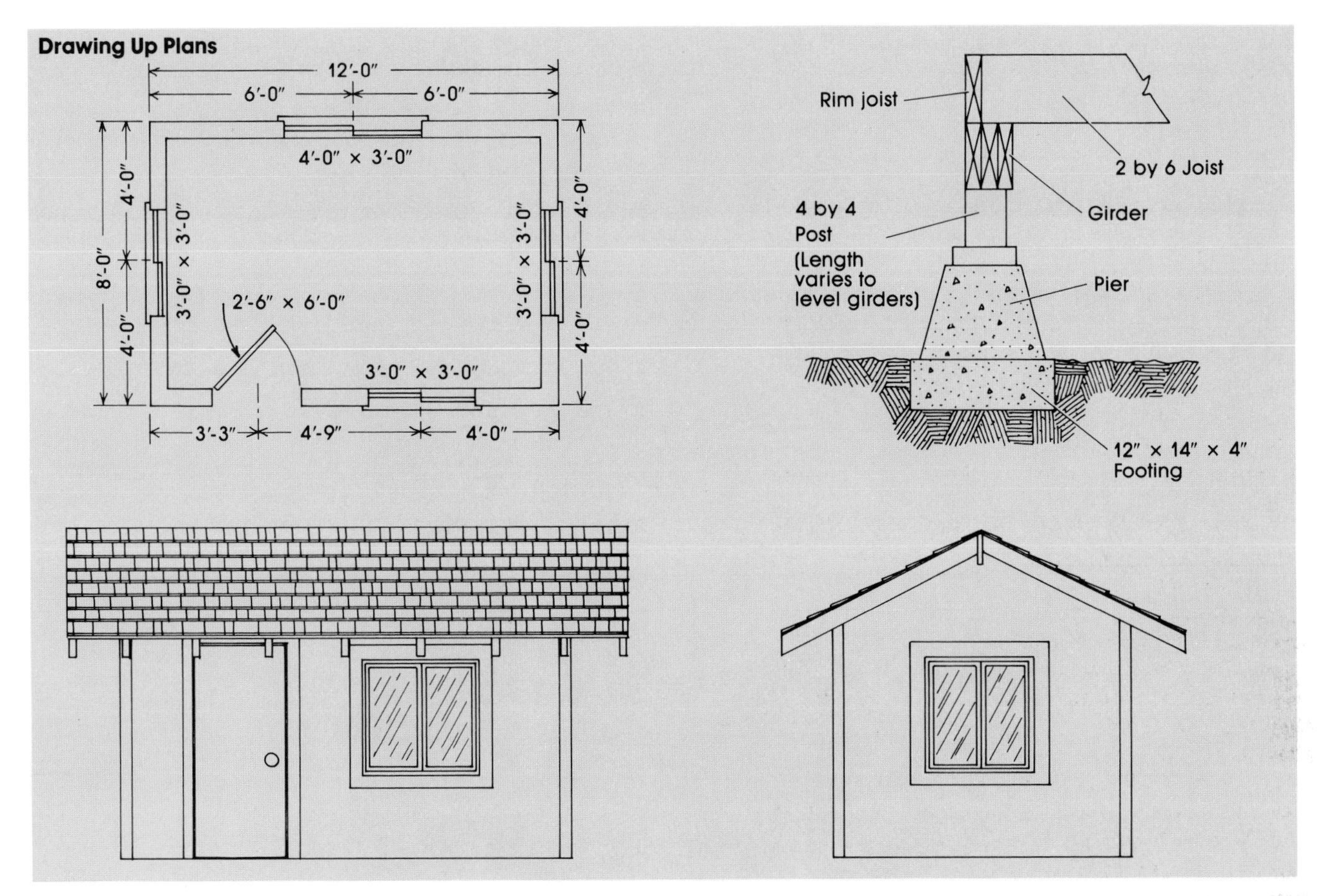


of your window space should be on the south side of the building so the low winter sun will provide solar heat. The north side, which will receive little or no sun in the winter, should have smaller windows and extra insulation. If you are going to have a deck or lawn around the cabin, allow room for it.

■ What types of trees surround the area? Deciduous trees can be quite beneficial, providing shade in the summer but letting the winter sunlight through when they shed their leaves. If you have fir trees nearby and you want lots of winter sun, you may have to relocate the house or remove a few trees.

■ Do you meet required setbacks? A setback, which is common in populated areas, is the distance the house must be set back from property lines. This is to prevent houses from being built too close together. If you are adding on a room or building a playhouse or guesthouse on a city lot, check with the building inspector's office for the required setbacks. You must know your exact property line to measure your setback distance. Failure to meet setback requirements could mean you have to tear down your structure.

## Permits and Building Codes

Even if you are building on your own property deep in a mountain retreat, you will probably need to get a building permit. The cabin you are planning to build will also have to conform to certain building codes. The closer you are to civilization, the more demanding these codes become. They are often irksome, but they are designed to make your house sturdy, weatherproof, and safe. They also enable your local government to reassess your property taxes according to the value of the improvements you make.

You can apply for a building permit as soon as you have a good set of plans, with several copies. The permit is obtained from the building inspector's office, usually located in city hall. A building inspector will study your plans to make sure they meet all local building codes and zoning regulations. Getting the permit can involve considerable red tape and delays, but at the same time the inspectors can be quite helpful to you. If you have questions about how deep the footings must be or what size rafters to use, they will have the answers. Often they are experienced builders themselves and can give you helpful and time-saving suggestions when you show them your plans.

How are you going to find out just what the minimum building code requirements are when you're making your plans? You can get a copy of the Uniform Building Code, but that is a large and complex book. You can consult with friends and neighbors who have a general understanding of construction. Or you can consult with a licensed contractor.

If you really want to build that cabin or room addition, but just aren't sure enough of your abilities, then hire an experienced construction carpenter or contractor and work with him. The work will progress much faster this way. Working with him, asking questions, and keeping your eyes open will give you an invaluable education.

Broken down into individual steps, the construction process is very straightforward. The steps for the playhouse/studio are outlined below and detailed on the following pages. You should have a similar outline and plan, as well as a detailed materials list, before beginning any project.

## Playhouse/Studio Construction Steps

**1.** Lay out the 8 by 12 foot building site. (See page 80).

**2.** Install footings for each of the 10 piers that are spaced every 4 feet around the perimeter of the building, and for the 2 piers down the center. Place piers on footings; align and level them.

**3.** Toenail 4 by 4 posts on top of piers. Cut all posts to the same level. Note that posts vary in length unless the ground is absolutely level.

**4.** Construct two 11-foot, 9-inch girders by nailing together three 2 by 6s for each. Construct the 11-foot, 9-inch center support girder from two 2 by 6s nailed together. Cut the two 8-foot end girders from 2 by 6s.

**5.** Put the side and end girders in place on top of the posts and piers. Once in place, be sure the foundation is level and square. Brace if necessary.

**6.** Cut 10 floor joists from 2 by 6s, each 7 feet, 9 inches long. Install every 16 inches on center. Align butts of one end 1½ inches in from the outer edge, the width of one 2 by 6. Toenail joists. Nail on two 12-foot 2 by 6 rim joists to cover joist ends.

**7.** Put down the subfloor of ½-inch shop grade plywood.

**8.** The walls of this playhouse are constructed from 6-foot studs to keep it in perspective for children but still high enough for adults to use since there is no ceiling inside.

Construct side wall A with one 4′0″ × 3′0″ window opening; bottom edge of all window headers is 5′9″ inches from the top of the sole plate. Raise side wall A into position and brace.

**Playhouse/Studio Construction Steps 1 thru 15**

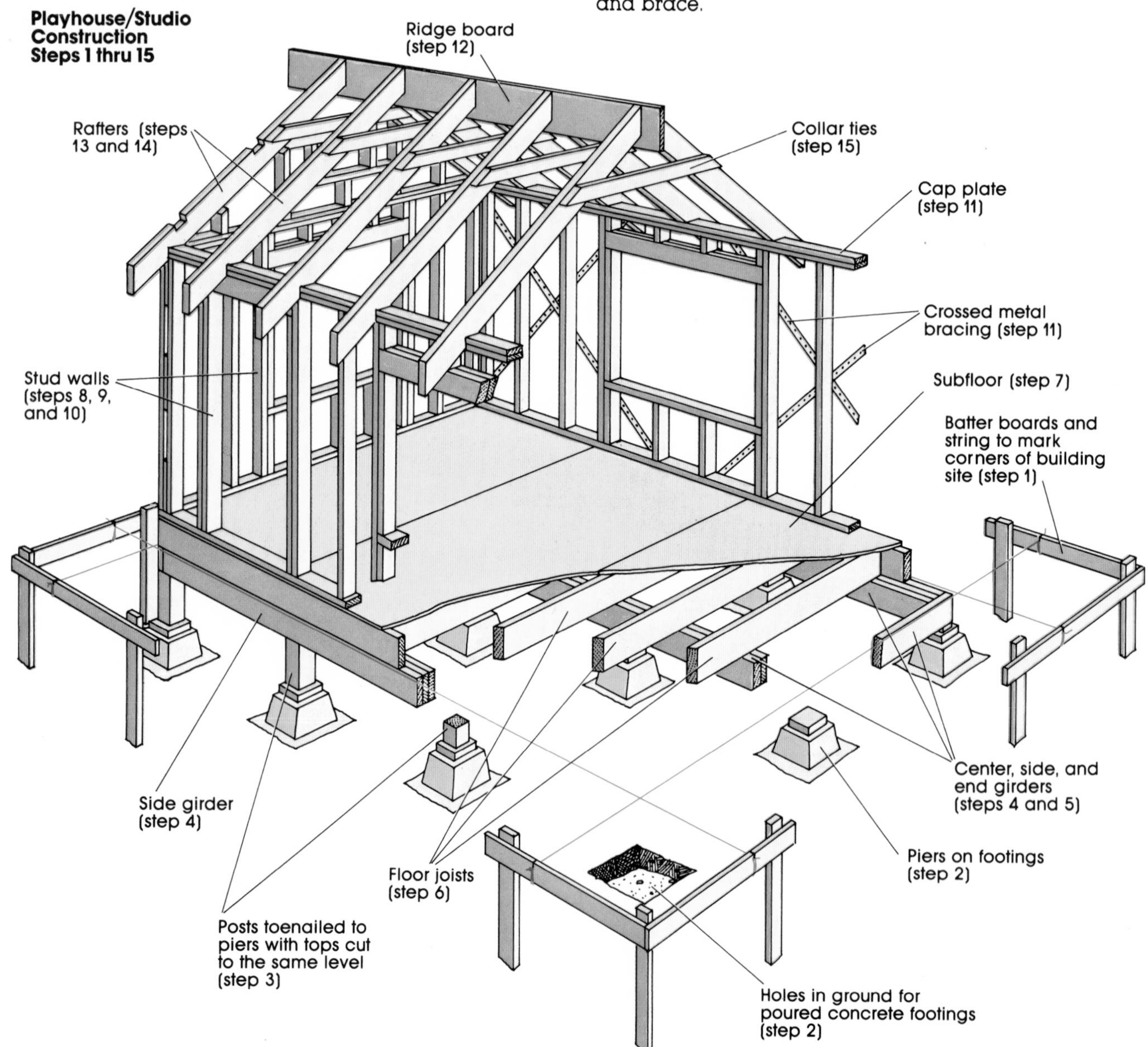

**9.** Construct side wall B. Locate door 20 inches from one end and center a 3′0″ by 3′0″ window opening 4 feet from other end. Raise and brace the wall in place.

**10.** Construct two end walls with 3′0″ × 3′0″ window openings centered in them. Raise and nail to side walls.

**11.** Plumb and square all walls. Be sure side walls are not bowing in or out. Nail on cap plates. Nail on crossed metal bracing.

**12.** Cut a 15-foot ridgeboard from a 1 by 6. Mark off the 18-inch gable overhang on each end, then mark off rafter locations on 2 foot centers. Put the ridgeboard up on temporary supports. For a 4 and 12 (1/6) pitch roof, ridgeboard is centered over end walls and is 19½ inches from top of cap plate to top of ridgeboard.

**13.** Cut two 2 by 4 rafters at a 4 and 12 pitch. Length of rafter to birdsmouth is 4′2⅝″. Put them in place to check your work, then use one as a pattern. Mark and cut 14 rafters with birdsmouth, plus 4 barge rafters without birdsmouth.

**14.** Erect all rafters and nail in place. Be sure end rafters are directly over end wall. Brace ridgeboard with 1 by 4s from center of ridgeboard to an end wall stud.

**15.** Nail 1 by 4 collar ties approximately one third down the length of the rafters.

**16.** Cut and put gable-end wall studs in place; include vent openings.

**17.** Cut and install outriggers for gable overhangs. Put on the 4 barge rafters, and nail in frieze blocks.

**18.** Put down starter board on gable overhangs and across rafter overhangs.

**19.** Nail on 1 by 4 sheathing across rafters. Space each one the width of a 1 by 4. Cover top 3 feet of roof solidly.

**20.** Staple 30-pound roofing felt over the roof and nail on shakes.

**21.** Install windows and door. (This door was made from a single sheet of ½-inch plywood cut to fit and then covered with scrap 1 by 6 starter board and stained.)

**22.** Put on siding and trim. Stain and paint.

**23.** Cover the interior floors with hardboard. Sheetrock the walls and cover the exposed rafters inside with starter board sheetrock or paneling for a cathedral ceiling.

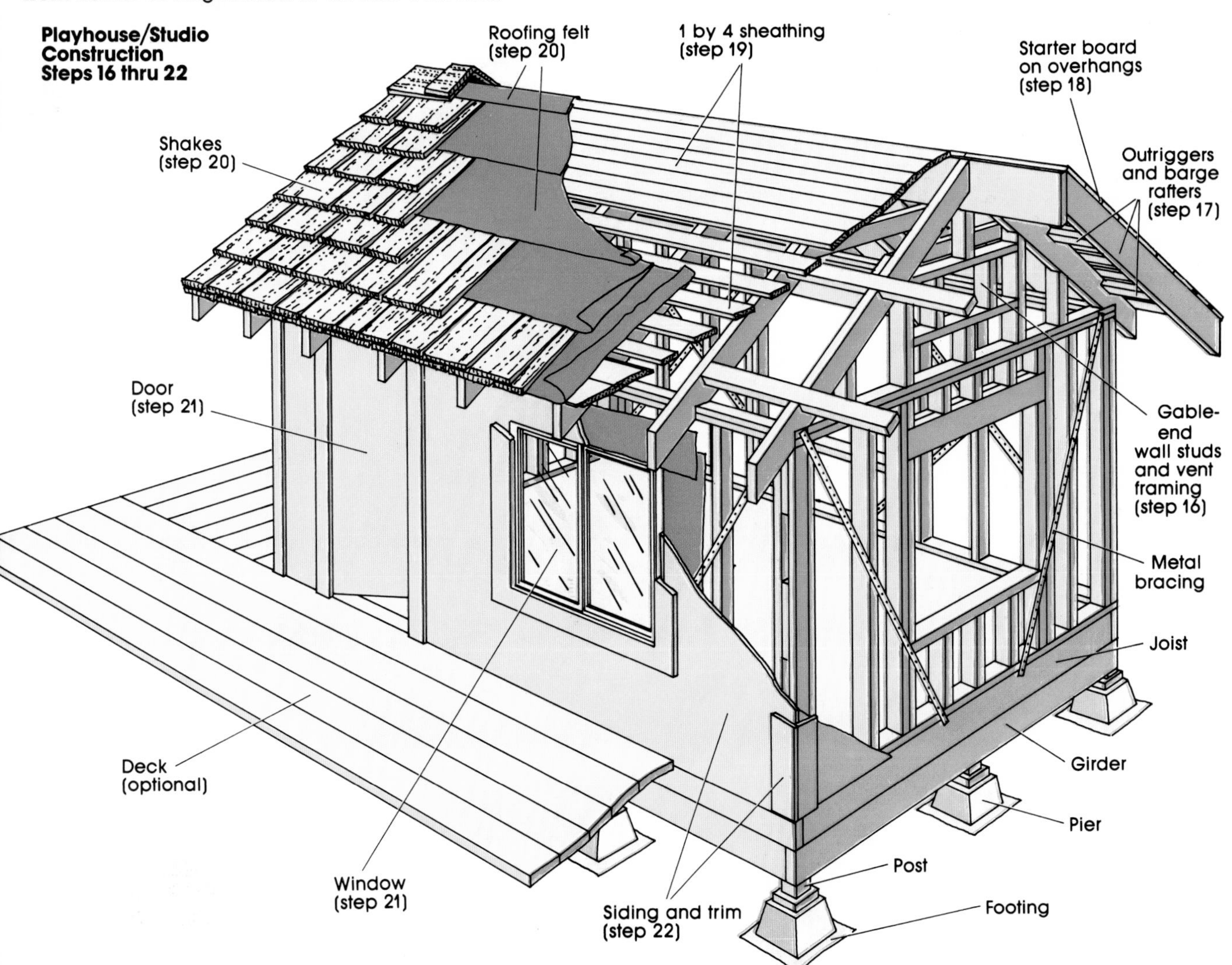

# LAYING OUT THE BUILDING SITE

The first step in actual construction is referred to as laying out the site. This means accurately locating the four corners of the building, which in turn will establish the boundaries of the foundation, no matter what type of foundation you use. The site is commonly laid out with the aid of batter boards. As shown in the illustration, the batter boards are set back from the corners of the planned building in an L -shaped arrangement. They allow you to maintain accuracy as you complete the footings and foundation walls.

Batter boards are made from stakes (2 by 4s sharpened with a saw or hatchet) that are connected with 4-foot lengths of 1 by 4. Each batter board should form a nearly perfect right angle. Use your framing square to check this. The tops of the batter boards should be level with each other all the way around. Stretch mason's twine tightly between them and check with a line level. (Mason's twine is a strong type of string that will stretch very tight.)

The basic steps in setting up the batter boards are as follows:

**1.** Determine where you want one corner of the building to be. Drive stake A firmly into the ground at that point.
**2.** Measure the long side of the building to where you want the next corner to be and drive stake B at that point. You have now established one long side of the house, and the next two corners will be determined from this.
**3.** Drive a small nail into the top center of each stake and connect them with tightly drawn mason's twine.
**4.** Locate the approximate spot for the third stake, C, by measuring and using your framing square for an approximate right angle. Connect stakes B and C with mason's twine, and set up stake D at the fourth corner in the same manner.
**5.** Now, using the batter boards, you must make the locations of the stakes exact. Erect the batter boards so that each corner stake is lined up directly on the diagonal from the opposite corner, as illustrated. Remember to check that the batter boards are all level with each other, using mason's twine and a line level.
**6.** Stretch a piece of mason's twine between the batter boards and over stakes A and B. When the twine is perfectly lined up, make a saw cut in the top of each batter board as a permanent mark. (You will have to remove the twine later.)
**7.** Stretch a piece of mason's twine between the batter boards over stakes B and C. This must form a perfect right angle with the twine you just stretched between stakes A and B. From stake B measure 6 feet back toward stake A and 8 feet toward stake C. Mark the two points with pins in the twine. You will have a true right angle when the diagonal distance from pin to pin is 10 feet. So with a friend or two helping, adjust the twine over the batter board behind point C until you have that 10 feet on the diagonal. Mark the batter board with a saw cut and tie the twine.
**8.** Continue this process until you have strings attached to the batter board at stake A. At each step, carefully measure from the point where the pieces of mason's twine cross to determine the exact length of each side of the building. Use a plumb bob dropped from inside of each twine intersection to locate the spot for each corner peg.
**9.** Double-check your work by measuring the two diagonals on the layout. If they both measure the same, the layout is square. If they don't measure the same, adjust the stakes until they do.

Once your layout is squared, you are ready to lay your foundation.

**Laying Out the Building Site**

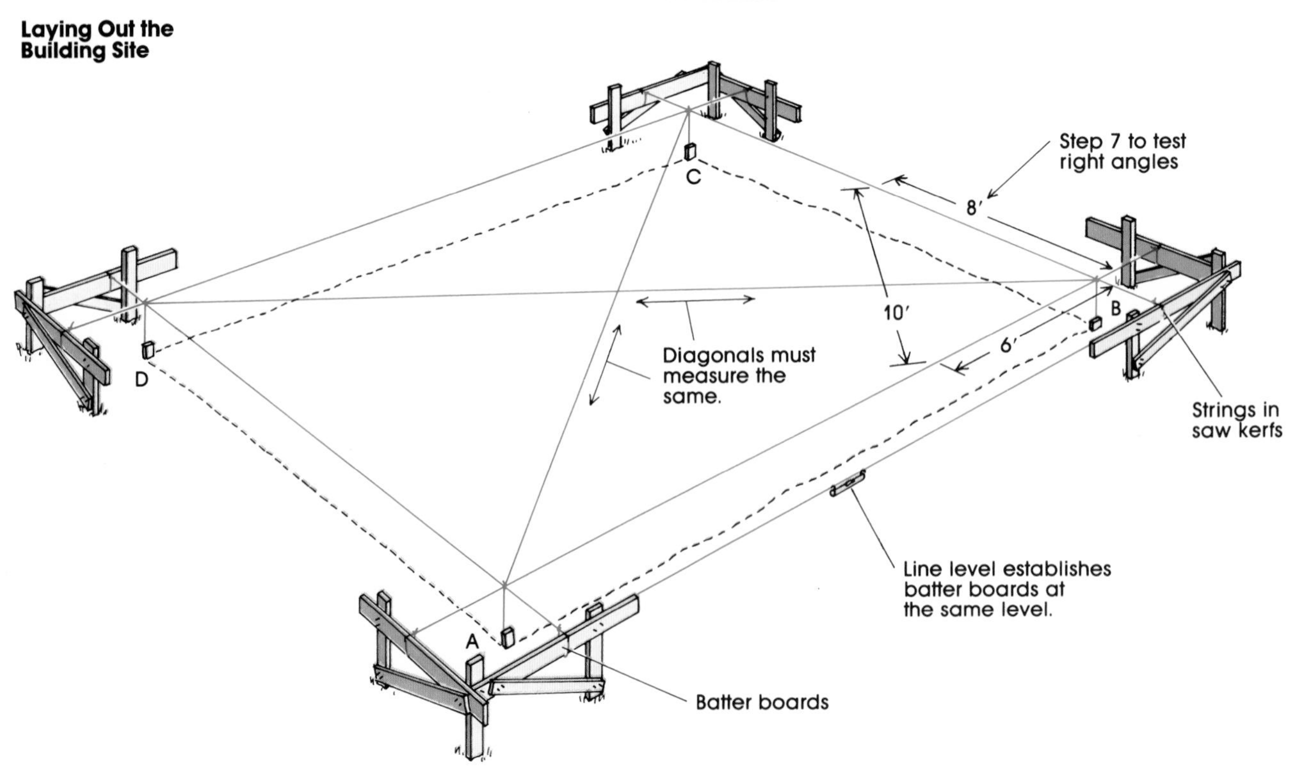

# FOUNDATIONS

Three types of foundations are covered in this section: the concrete block wall, the post and pier foundation, and the concrete slab.

The concrete block foundation requires footings, which are placed in a trench and on which the foundation wall rests. This type of foundation is the most difficult of the three to construct, but it is also one of the most permanent. The foundation wall raises the house away from the earth, which is often damp and may harbor termites and other insects.

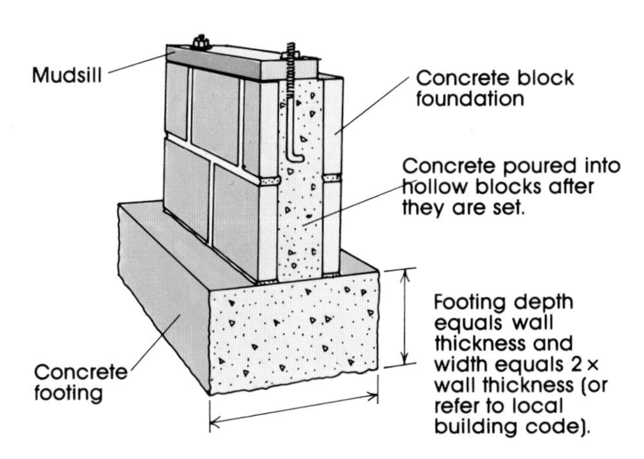


The post and pier foundation may require footings under each pier if it is built on a slope or if local codes require them. However, footings are often unnecessary. Even including footings, this is one of the easiest and cheapest types of foundations to construct. It is ideal for vacation cabins where supplies must be hauled a considerable distance.

The concrete slab foundation rests directly on level ground (or on a bed of gravel if codes require it). It is strong, long-lasting, and fairly easy to install. It is an excellent choice for a garage, shop, or small barn.

Another type of foundation, called the stem wall, is widely used by professionals but is too complex to be described in detail in this book. The stem wall involves digging the footings trench, building a form wall above the trench to the proper foundation height, and then pouring the footing and foundation wall in a single step.

## Concrete Block Foundation

### Footings

When your layout is squared, you are ready to prepare the footings that support the foundation wall and the entire house. A quick call to the building inspector's office will tell you how deep you must go in your area. However, the rule of thumb on footings is to put them on solid earth (not fill dirt) below the frost line. If the footings are above the frost line in cold climates, the freezing earth will heave the footings upwards during the winter months, causing your house to shift slightly. This results in sticking doors and cracking plaster, and sometimes entire subfloors pull apart.

Footings should be as thick as the foundation wall and twice as wide. In other words, if your foundation wall is made of 8-inch-wide concrete blocks (which is standard), the footing should be 8 inches thick and 16 inches wide, as shown in the illustration on this page. Again, check your local codes first.

Before going further, note that even houses with foundation walls around the perimeter generally use posts and piers down the center to support the center beam for the floor joists. See the post and pier section for details on this.

While footings should be put down carefully, they need not be absolutely precise because the foundation wall placed on top of them can be adjusted slightly.

**Laying out footing lines.** To lay out the footing lines for a 16-inch wide footing, make saw cuts in the batter boards 4 inches to the outside and 12 inches to the inside of the perimeter twine. This way, the 8-inch-thick foundation block will be centered on the footing, and the outside edge will be directly under the perimeter line.

Now pull new twine lines over these cuts all the way around. At each intersection, drop a plumb bob and drive a small stake. Connect all the stakes with twine and you have the inside and outside lines for your footing trench. Once you have marked an outline in the dirt with chalk or a pick, remove all the twine so you can begin digging.

The amateur home builder normally digs footing trenches by hand. Use a pick and a square-tipped shovel to keep the trench walls straight and the bottom as flat and as level as possible.

**Trench footings and leveling stakes.** A simple but effective method of putting down the footings for a cabin or barn is to pour concrete directly into the trench, using the trench for a mold instead of building forms. The trick is to make the footings perfectly level.

Once you have completed the trench, check that it is reasonably level by resting your level on a long, straight 2 by 4 placed on the bottom. Then drive a 2-by-2 stake into the ground in the center of the trench at one corner until the part of the stake above ground is the prescribed thickness of the footing. Measure off 6 feet down the trench and drive another stake. Check that its top is level with the

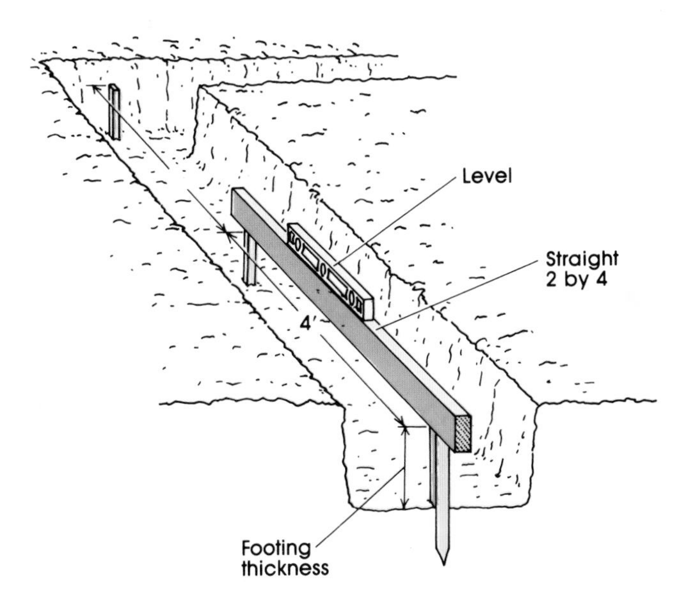

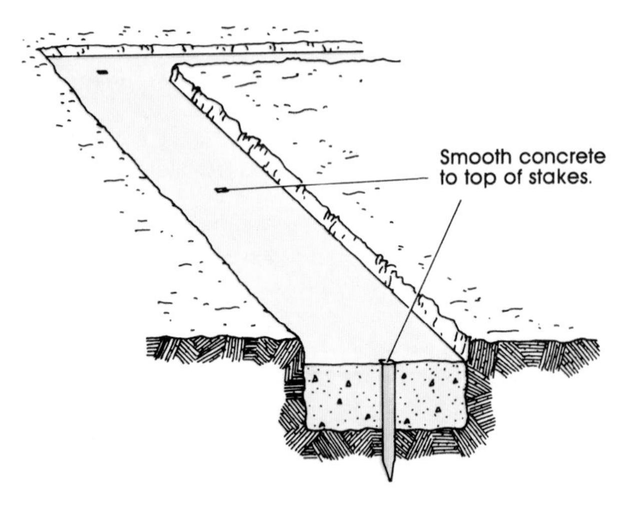


first one by placing a level on top of a straight 2 by 4 resting on both stakes. Continue around the trench in this manner.

A trick to ensure that you are staying level is to reverse the 2 by 4 and level for each measurement. Any slight variation in the bubble will be more noticeable this way.

When the stakes are all level, fill the trench with concrete just to the top of the stakes. Work the concrete with your shovel to remove any air bubbles, then trowel it flat and smooth. Leave the stakes in place.

**Footing forms.** A more precise means of squaring and leveling the footings is to use forms. The footings ditch must be considerably wider than 16 inches to allow you room to work—about 30 inches will do.

Once the ditch is finished, start the forms at one of the outside corners. Find the corner by dropping a plumb bob down from the twine and driving a 1-by-2 stake there.

The forms can be made from either 1-by or 2-by material, the latter being more expensive but requiring less bracing.

For an 8-inch-thick footing, use a 1-by-8 board and keep it about a ½ inch off the ground. Place the boards for the outside perimeter first, nailing them to stakes placed on the outside of the boards about every 18 inches. Also place a stake on each side of the joint where two boards meet.

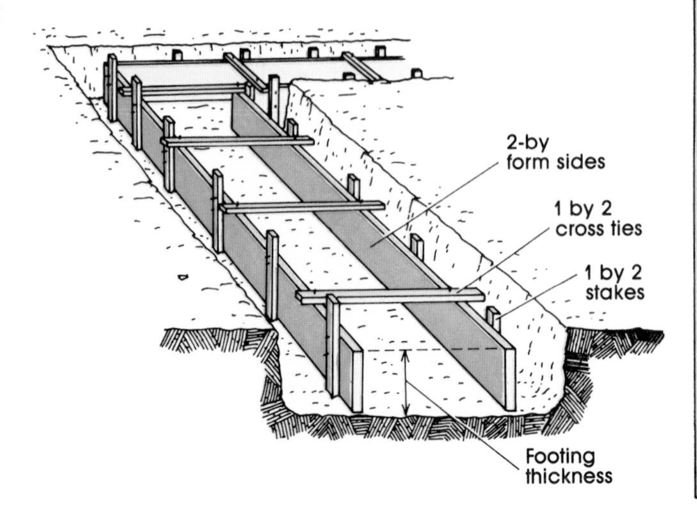


Cut several 1 by 2s or pieces of scrap 16 inches long to use as spacers, and put up the inside boards in the same fashion as the outside boards. When the form is complete, nail 1-by-2 crossties across the top of the form every 2 feet to keep it from spreading under the weight of the concrete.

Pour the concrete flush with the top, work it with a shovel to settle it, and then trowel it smooth, cover with plastic, and let it cure. You can begin putting down the concrete block foundation wall the next morning, even though the footing may not cure completely for five or six days.

### The Foundation Wall

When the footings are complete, you are ready to put up the foundation wall. This wall is designed to raise a building far enough above the ground so it will not be affected by soil moisture or wood-boring insects. It is generally made from masonry blocks. Blocks are nominally 6, 8, or 10 inches wide, 8 inches high, and 16 inches long. The actual measurements are about ⅜ inch less, which allows for the mortar at each joint. An 8-inch block is standard for a foundation wall.

Mortar is concrete without the gravel. To mix mortar for the blocks, use 1 part masonry cement and 3 parts fine

## Mixing Concrete

To mix your own concrete in a wheelbarrow (which is very laborious) or cement mixer, use 1 part portland cement, 2 parts sand, and 3 parts gravel. These ingredients are commonly measured by the shovelful.

The other ingredient, of course, is water. Allow 5 gallons of water for each sack of cement. This works out to about 3 quarts of water to 1 shovelful of sand. You can adjust these rough proportions according to the amount of water in your sand, which is normally sold wet. Squeeze a handful of sand, then open your hand. If the sand crumbles, it is dry; if it compacts smoothly, it is average; and if your hand is visibly wet, the sand is very wet. A little experience will allow you to eyeball the amount of water. The mix should be an even gray color, creamy and almost plastic in texture. It should not be soupy or dry and crumbly. The trick is to add water carefully and slowly as you approach what appears to be the right mix.

If you mix the cement in a wheelbarrow, work the sand and cement together with the shovel before adding the gravel, then turn everything a few times before adding the water. Add the water slowly and turn the mix repeatedly with the shovel until it has the right color and texture. Use a flat-end shovel so you don't leave pockets of cement in the corners of the wheelbarrow.

Footings should be covered with plastic to retain moisture during the curing process. Curing time depends on air temperatures, but three days in warm weather or five days in cool weather is generally enough. Although the concrete may not be fully cured, you can almost always work on it the day after it is poured. If the temperature is below freezing, seek special advice from a local concrete-mixing firm.

sand. Since it dries out fairly quickly, don't mix more than half a wheelbarrowful at a time.

The first step is to lay out all the blocks on top of the footings, spacing each one with a scrap of 3/8-inch plywood to make sure everything fits. Then take all the blocks off the footings and restring the building perimeter line as you did for the footings. Locate the outside corner of the corner blocks with a plumb bob.

Spread a layer of mortar, commonly called mud, along the footing for the space of one or two blocks. Place the corner block firmly in the mud so there is just a 3/8-inch layer between it and the footing. Use the end of a trowel handle to tap it square and level. In the same manner, put another block at the other end of the footing. Now you are ready to lay the blocks in between.

Note that the web in these hollow-core blocks is narrow on one side and thick on the other. Always lay them with the thick side up.

Stretch a piece of mason's twine along the top outside edges of the two end blocks. This is the guide to keep your wall straight and level.

When you come to the last block in a course, butter both ends with mortar and slide it carefully into place. If the mortar falls off, pull it out and do it again. Continue around the footing in this manner.

Most codes call for the hollow spaces in load-bearing walls to be filled with concrete (see illustration on this page). Some codes even call for reinforcing rods (rebars), so double-check.

Even if the blocks don't need to be completely filled with concrete, you will need to fill holes every 4 feet to place the anchor bolts, which will hold down the 2 by 4s that serve as the bottom plate for the stud wall. Let 2½ inches of each bolt protrude above the concrete, and space the bolts so none of them will be placed at the 16-inch intervals where the studs will fall. Since studs start 16 inches from the end of a wall, put the first anchor bolt 12 inches from the end and space all others 6 feet from that.

With the anchor bolts in place, you're ready to put the mudsill down and then start framing.

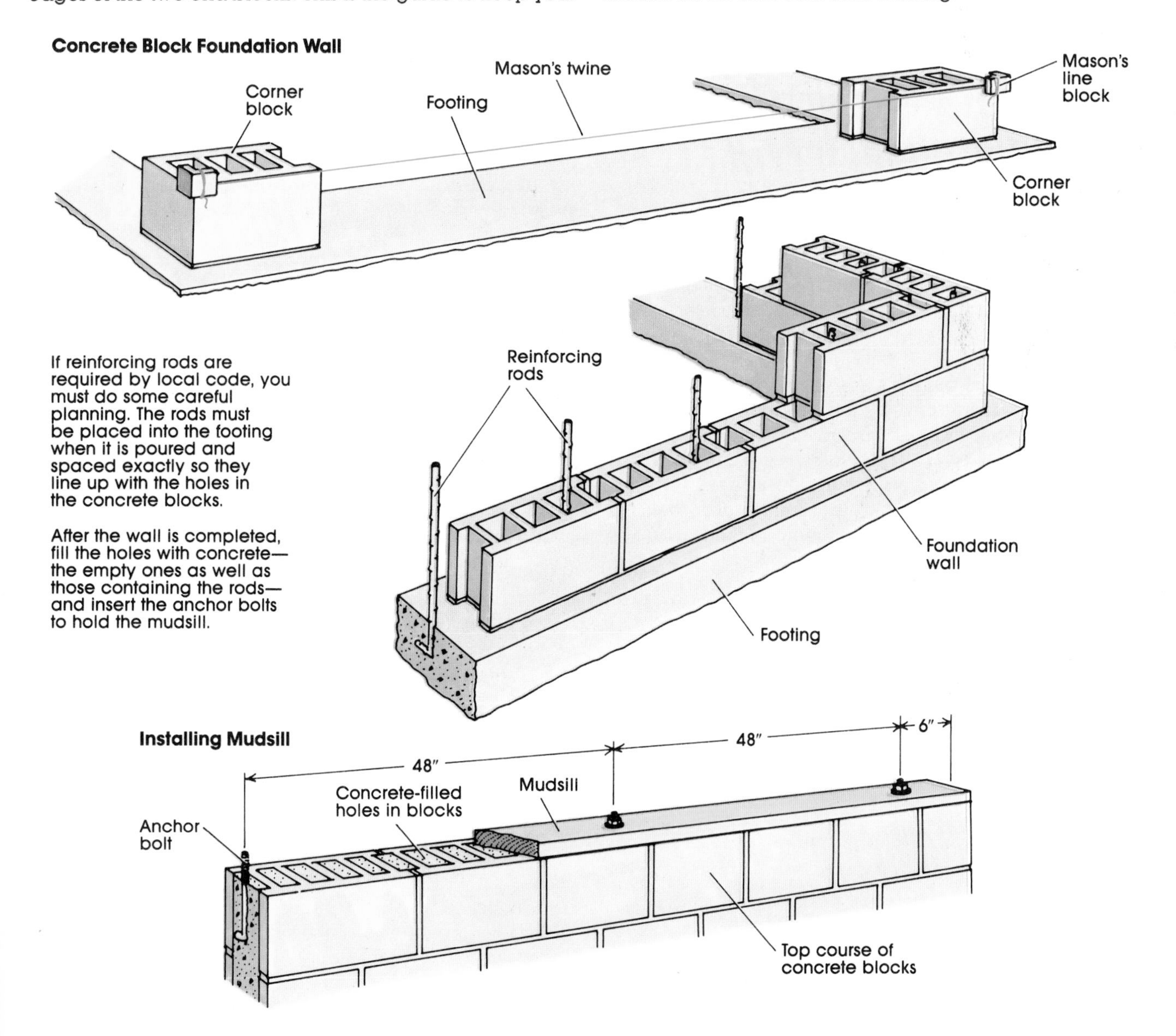

### Post and Pier Foundation

Piers are precast tapered concrete blocks that are 12 inches square at the bottom and have a 6-inch-square piece of redwood set in the top. Because they have a wide base, they sometimes do not require footings underneath, but you should check your local codes. Piers are placed in position on the ground or on footings to provide support for posts. The posts form a level base for girders, which function as foundation walls. If you are building on sloping ground, it's advisable to dig footing holes 6 to 12 inches deep and fill them with concrete. Set the piers in place while the concrete is still wet to prevent them from slipping. Place a level on top to ensure that they are level, both north-south and east-west.

**Pre-cast Pier**

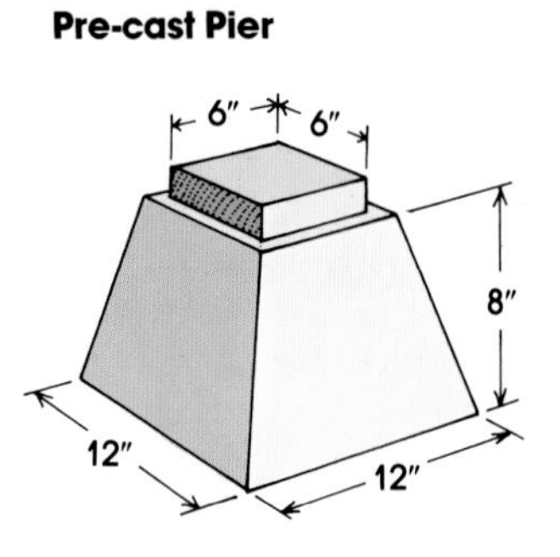


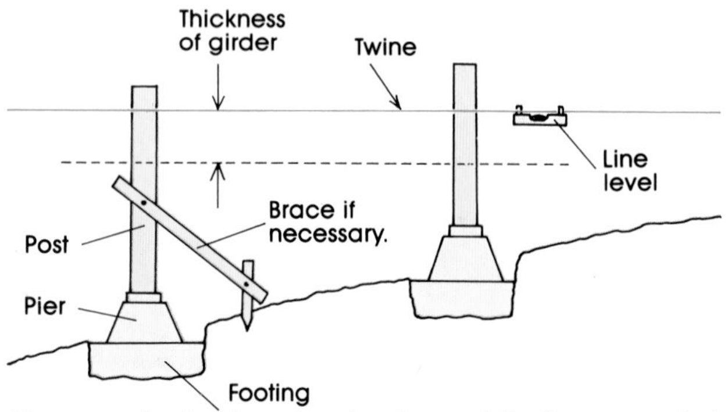


The number of piers you need and the distance between them is determined by the size of the girders. See page 87 for a discussion of girders and refer to the span charts on pages 108–110 to decide what you need.

Once your layout is squared (see page 80), use a plumb bob to set the piers so the outside corner of the redwood top falls directly under the tip of the bob. Once the four corners are set, align the intervening piers along the twine stretched between the corner piers.

The piers do not have to be all on the same level. These differences in height are corrected by cutting 4-by-4 (or 6-by-6) posts so their tops are all on the same level. The perimeter girders are then nailed to the top of the posts to establish the level and square floor foundation.

To find the heights of the posts, start at the highest pier and stretch twine to the next corner. Use a line level to check that the twine is level, then measure from the twine to the tops of the piers. Cut the posts to the proper heights and toenail them to the piers. Check your accuracy by placing a straight 2 by 4 across the tops of the posts and checking it with a level.

On fairly level ground, posts may range in height from a few inches to a foot or two. If they are much higher, you may have to brace them with 1 by 4s. When the posts are up, place the girders on top of the posts. Girders are commonly made of two or three 2 by 6s nailed together. When the girders are in place, you are ready to put down floor joists and start framing. (See page 87 for details on construction and placement of girders).

### Interior Posts and Piers

Whether you use a perimeter wall foundation or all posts and piers, you will probably need one or more rows of piers in the interior to support the middle of the house. Again, you can determine the number and spacing of these piers according to the size of the building. Use the span charts on pages 108–110.

Once the exterior walls or girders are in place, align the center piers by stretching twine from end wall to end wall. Set the centers of the piers directly under the twine, checking that they are level and on firm ground.

You are now ready to measure the heights of the posts for the center support girder. While measuring, remember that the twine stretched across the top of the foundation wall represents the *top* of the girder, so measure from the twine to the tops of the piers and then subtract the width of the girder to find the heights of the posts.

When the center girder is in place, you are ready to put down the floor joists.

**Corner and Interior Posts and Piers**

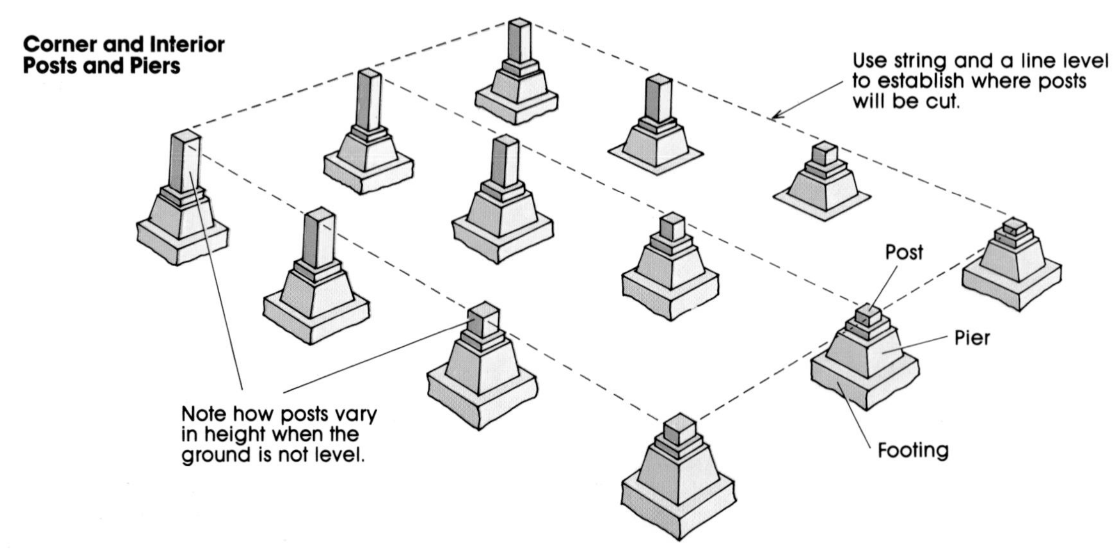

## Metal Hangers, Post Supports, and Braces

When undertaking large-scale construction, such as houses, decks, or barns, you can use a variety of steel supports to tie beams together, hang joists, support posts, or brace walls. Hangers and post supports come in different sizes according to the size of the joist beam or post they will support. Most hardware stores carry a wide selection.

Some of the most widely used hangers, supports, and braces are illustrated here.

**Joist hanger.** This metal hanger is designed to support a joist or a beam. It is nailed to the face of a crossmember, and the joist or beam slips into it. Joist hangers range in size from 2 by 4 to 4 by 10 inches. Larger ones can be specially ordered.

**Post supports.** Designed to provide a base and support for posts, such as under decks, these supports are commonly available for 4-by-4 or 6-by-6 posts. Larger ones can be specially ordered. They are made in two basic styles: fixed and adjustable.

The fixed post support has metal flanges down the sides that are set into fresh concrete. The post is then set in place and nails or screws are driven through the side supports.

The adjustable post support is primarily used on existing concrete slabs. A hole is drilled in the concrete and the support is fastened to the slab with an expansion bolt (see page 51). The post is set in place and nails or screws are driven through the side support. One type of adjustable post support has an inverted T-shaped steel plate that rests between the support and the post base to keep it off the slab and away from possible moisture.

**Post cap.** This is used to tie a horizontal beam securely to the top of a post. It comes in two pieces, one for each side of the connection. Post caps are available in a variety of sizes for 4-by-4 and larger posts.

**T-strap.** This simple version of the post cap consists of a steel T with nail holes in it. The top of the T is nailed to the beam and the bottom of the T to the post. It is also used to tie two beams together where they meet in the center of a post.

**Corner post cap.** This is similar in all respects to the standard post cap except that one part of the T-wing is missing, since the beam does not extend beyond the post.

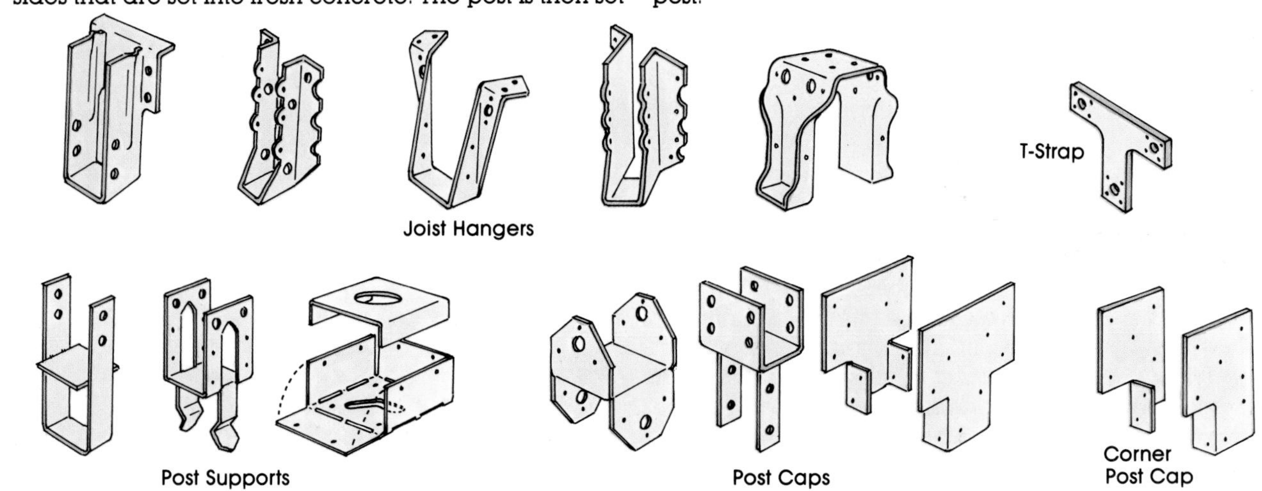


## Concrete Slab Foundation

After piers, concrete slab foundations are among the easiest to put down from a technical standpoint. It's a tough job, however, in terms of the physical labor involved. The only realistic way to put down a slab is to have a ready-mix truck deliver the concrete for you.

The instructions provided here are for putting down a slab for a project such as a barn or a workshop. If you want to put down a slab foundation for a house, you must place all the plumbing lines first. Since plumbing and wiring are not covered in this book, you should get professional assistance with these aspects of building a house before pouring the foundation.

Building codes in colder parts of the country may require you to dig a footing trench around the perimeter of the slab to support the edges. Codes in milder areas generally do not require this. The thickness of the slab is also determined by local codes, but 3½ inches—the width of a 2 by 4—is generally the minimum requirement. Some codes also require a bedding of gravel several inches thick under the concrete slab. In any case, the ground must be very nearly level before you can begin laying the slab.

### Laying the Slab

**1.** Locate the corners using batter boards (see page 80 for details).

**2.** Create a form for the concrete by cutting 2 by 4s to length for the slab perimeter. If the slab is going to be larger than 12 by 12 feet, divide the area into more easily manageable sections by adding 2-by-4 form boards to the interior of the slab area. You can remove these boards when the section hardens if you want a smooth slab, or you can leave them in place. Use redwood or treated lumber if you want to leave them in place.

**3.** In addition to being square, the perimeter form boards must be level. Space 1-by-2 stakes every 3 feet along the outside, and nail them to each board while your partner

**Concrete Slab Foundation**

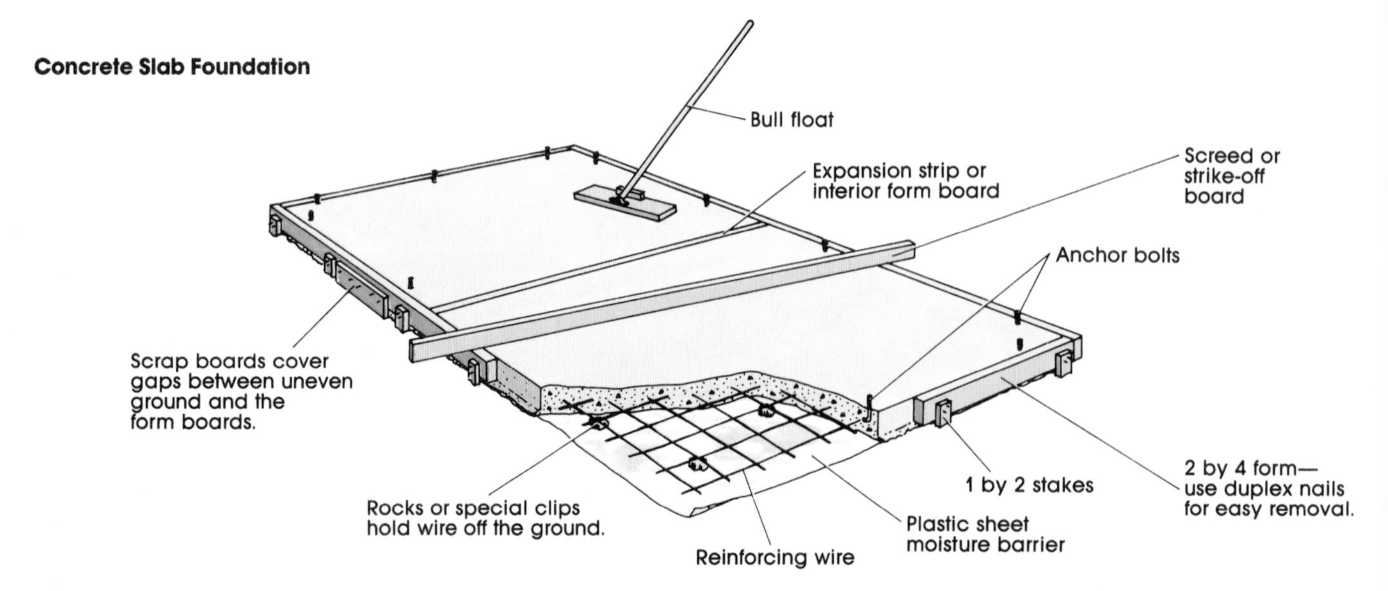


holds it level. Work your way around the perimeter in this fashion, constantly checking the level. The form boards should be within 2 or 3 inches of the ground. Fill any gaps larger than this with scrap wood.

Stakes can be tapped in or worked up for small adjustments. Make sure the tops of the stakes are all below the top of the form so you won't run into them when you are leveling the concrete. Check that everything is square by measuring the diagonals of the completed form.

**4.** Cover the interior space with a layer of 4-mil plastic sheeting to form a moisture barrier. This prevents any groundwater from working its way through the floor.

**5.** Now cover the interior with 6-inch-square reinforcing wire. This is particularly important if you will be placing heavy equipment such as trucks or tractors on the floor, because the wire runs through the center of the slab to bind and reinforce it. For example, the wire should be raised off the ground about 2 inches for a 4-inch slab, so that it will be in the center of the slab's thickness. You can do this by placing a series of small rocks every 3 or 4 feet under the wire to hold it up. Some hardware stores also carry special metal clips that fasten to the wire and hold it off the ground.

**6.** When the ready-mix truck arrives, be ready with several workers. Concrete firms allow only a few minutes to unload each cubic yard of concrete and then start charging you waiting time.

As the concrete is sent down the chute, spread it with shovels as evenly as possible. Bring it up about ½ inch above the form boards, since it will settle. Work it lightly with a shovel to help it settle.

**7.** Now, with a straight 2 by 4 long enough to reach from form to form, "strike off" or level the fresh concrete. This is done by moving the board across the concrete in a sawing motion toward you.

**8.** Go over the top with a bull float, which works the gravel down from the surface and brings a watery sheen to the top. Whenever you push or pull, raise the advancing edge slightly so you don't push the concrete.

**9.** Once you have bull floated, press anchor bolts into the concrete. These bolts will hold down the 2-by-4 redwood sill or treated lumber that serves as the bottom plate for the stud wall (see page 90). Set the bolts ¾ inch back from the edge of the slab. In order to prevent them from being placed at the 16-inch intervals where the studs will fall, place one bolt 6 inches from each corner and space the other bolts at 4-foot intervals starting from that point. Don't put any anchor bolts in door openings.

**10.** When the watery sheen has disappeared, run a mortar trowel around all the edges to separate the forms from the concrete. Use an edger next to give the edges a finished look.

**11.** For barns or shops, you can help prevent the slab from cracking by expansion as a result of heat or cold by installing control joints every 10 feet. (Control joints are not necessary for a house.) Use a grooving trowel to cut these joints. Guide it with a long, straight board, as illustrated. Be sure to put down a 2-by-6 catwalk or knee boards to disperse your weight on the soft concrete.

**12.** Immediately after this, finish the entire surface the way you want. Running a push broom back and forth will give you a good nonslip surface. Use a wooden trowel for a smoother finish. A steel trowel will give you a near-polished surface that is unnecessarily slick for a work area.

**13.** To cure, cover the concrete with 4-mil plastic sheeting and weight down the edges to trap all the moisture. If you use burlap or straw instead of plastic sheeting, keep it constantly damp. The slower the cure, the stronger the concrete. Keep the slab wet or covered for at least five days.

**Trowels**

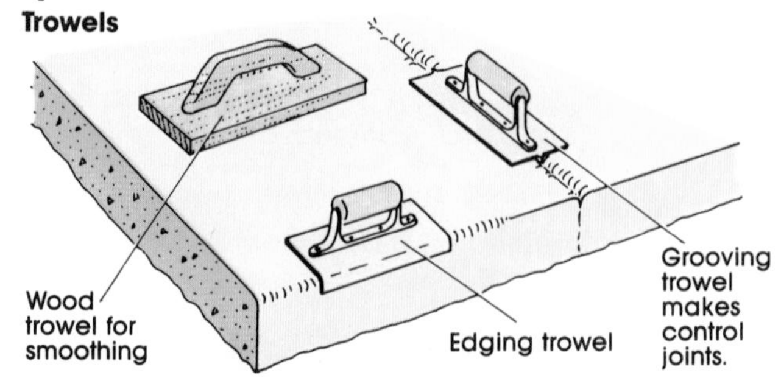


The concrete will be hard enough to work on the following morning, although it will not be cured. You are now ready to put down the mudsill and begin framing.

# GIRDERS, SILLS, & FLOOR JOISTS

Three types of foundations have been discussed so far: the concrete slab, the concrete block wall, and the post and pier foundation. The concrete slab requires a sill fastened to the concrete; the concrete block wall requires a sill fastened to the concrete and a support girder down the middle on posts and piers; and the post and pier foundation requires wooden girders that rest on the posts. This page demonstrates how to put the sills and girders in place. When the sills or girders are complete, you're ready to put down the floor joists.

## Girders

In post and pier construction, girders span the posts and piers to form the building perimeter. They run down the two long sides of the building and down the middle. The ends of the girders are tied together across the width of the building with single 2-by joists the same width as the girders. Single joists rather than built-up girders are used across the width of the building because they are non-load-bearing.

Before installing the girders, you should cut the posts to length so the girders will all be level. Detailed instructions for doing this are provided on page 84. The girders should measure 3 inches less than the length of your building to make room for the two end joists. They should be as long as possible, but if your building is longer than 20 feet you will have to make two of equal length to meet on a post in the middle.

Construct each girder from three lengths of 2 by 6 or 2 by 8 (check the span chart on pages 108–110). To make the girders, nail together two boards, then nail on the third. Space 16d nails every 18 inches and stagger them from side to side.

Put the girder up, brace it in place, and check your work with a square. Make any adjustments necessary. If two girders meet on a single post, tie them together with metal straps (see illustration) or with 2-foot lengths of plywood.

Once the basic girder framework is up, measure the diagonals again. It's very important that the girders be within ½ inch of being perfectly square. If they are off by any more than that, you will run into trouble when the wall framing starts.

## Sills

The sill, commonly called the mudsill, is usually made with 2-by-8 redwood or preservative-treated wood. (A 2 by 8 completely covers the top of an 8-inch-wide concrete block.) It is anchored to the foundation with anchor bolts and is designed to prevent the building from being shaken off the foundation by earthquakes or high winds. It also provides the nailing surface for attaching the floor joists.

Lumber for sills should be carefully selected for straightness. If the building is too long for a single sill (20 feet is the maximum length you can buy), then plan accordingly when you place the anchor bolts (see page 83).

The first step in installing a sill is to hold each 2 by 8 on top of the anchor bolts. Use a square to check that the ends and sides are precisely flush with the foundation edge. Then rap the board sharply with a hammer over each anchor bolt to mark its location.

Next, drill holes for the anchor bolts that are ¼ inch larger than the bolts to allow for small adjustments. Slip the sills over the bolts and snug them up lightly using large washers and nuts. Then measure the diagonals to check that the sills are square.

## Floor Joists

Joists provide the main support for flooring. They are commonly made with 2 by 8s or 2 by 10s for maximum span. Joists for a concrete block foundation wall rest on the sill at each side and on the girder down the middle. Joists for a post and pier foundation rest directly on the girders, and a sill is not necessary.

Joists should always be placed with the crown, or high point, up. Determine the crown by sighting down the edge

**Floor Joists**

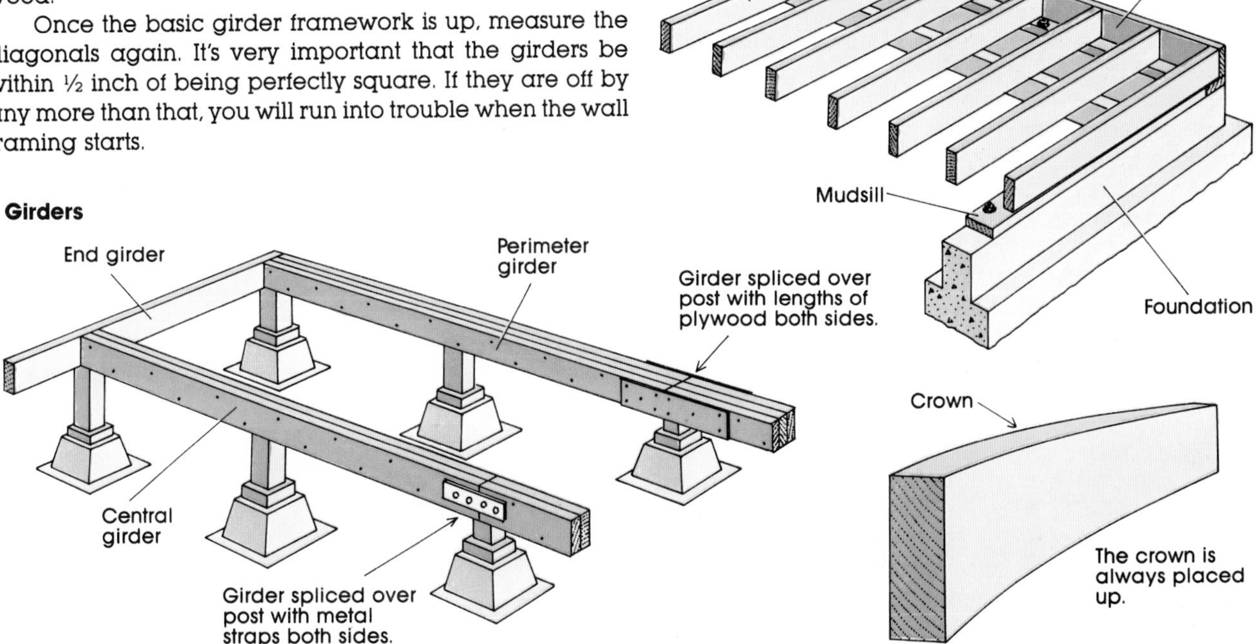

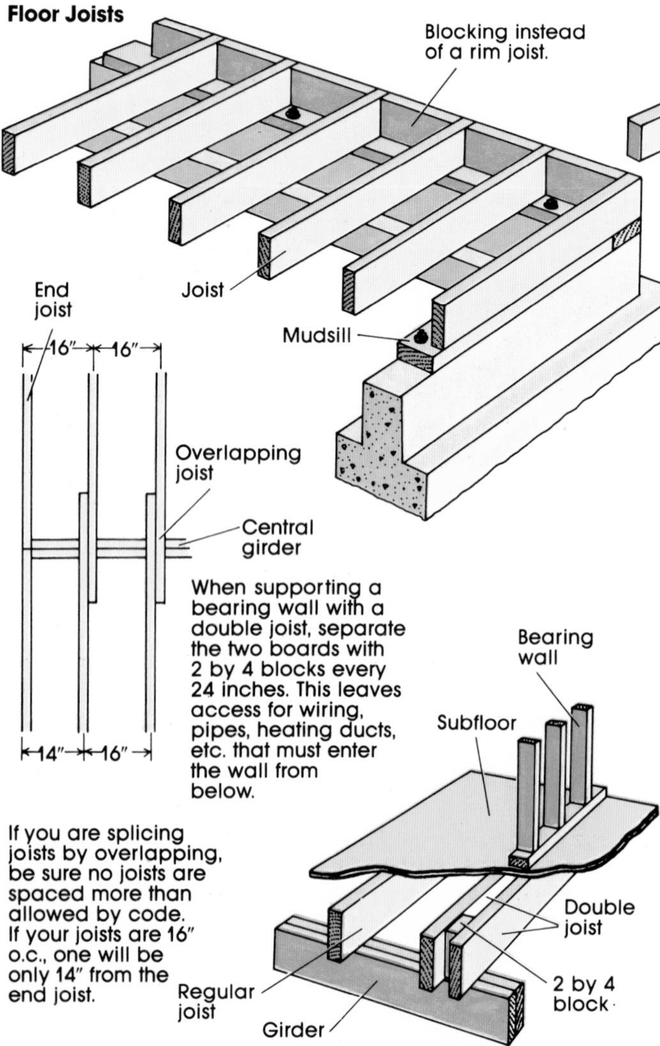


When supporting a bearing wall with a double joist, separate the two boards with 2 by 4 blocks every 24 inches. This leaves access for wiring, pipes, heating ducts, etc. that must enter the wall from below.

If you are splicing joists by overlapping, be sure no joists are spaced more than allowed by code. If your joists are 16" o.c., one will be only 14" from the end joist.

Joist spliced by overlapping on central girder.

Central girder

Joist spliced with plywood on both sides.

of each board before placing it. The weight of the floor will tend to straighten the crowns. Any large knots should also be on the upper side of the joist for maximum strength.

Joists do not necessarily run the full width of the building. When they are in place, the ends are often capped with a board called a rim joist that is the same stock as the joists. Thus, the joists are measured and cut 3 inches less than the building width to allow for the two 1½-inch rim joists.

An alternative method is to run the joists the full width of the building and then cut blocks to fill the spaces between them.

The first or end joist should be flush with the end of the building. The remaining joists are placed 16 inches apart on center (on center or o.c. means that the center of the joist goes over the mark). Put down a double joist wherever you plan to have load-bearing interior walls (walls that support ceiling joists).

To align the ends of the joists, snap a chalkline down the length of the sill 1½ inches in from the outside edge. If you are using girders, the joists should cover the inside two of the three boards that make up the girder. Toenail the joists to the sills or girders with two 16d nails on each side. Nail on the rim joists with three 16d nails in each joist end.

### Joists that Span Half the Building

If your building is more than 16 feet wide, it's best if the joists run only to the center support girder. Boards come in standard lengths only up to 20 feet, and at that length there may be considerable warping. Besides, they are hard to handle by yourself.

To make your work a little easier, use boards that run the full width only for the two end joists, if you can find ones long enough. Otherwise, butt two of them together at the center of the building and toenail them in place.

The remaining joists can be overlapped across the center girder, as shown in the illustration. Alternatively, you can butt the joists together on the center girder and splice them together with 2-foot lengths of plywood or 1-by lumber.

### Blocking

Blocks made from the same stock used for the joists should be placed between the joists down the center girder. This adds stability and helps prevent any twisting as the joists dry. Stagger each block for easier nailing.

If the joist span is more than 12 feet, staggered blocking down the midpoint of the joist span also helps prevent twisting.

Once the joists are in place, you should run all the plumbing lines, which are commonly fastened to the floor joists. When the rough plumbing is complete, you are ready to put down the subfloor.

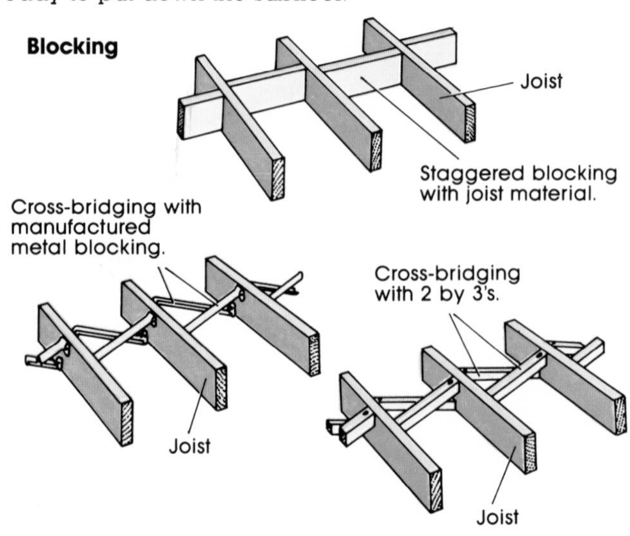

# SUBFLOORING

Subflooring gives stability to the foundation and provides a platform on which to frame the house. When the subfloor is finished, you can consider a one-story building about half-framed.

There are two ways to install the subfloor: with 1-by-6 boards laid straight or at a 45 degree angle, or with plywood sheets. The only reason to choose 1 by 6s over plywood for subflooring is if you can find a particularly good deal on the price of the boards. Plywood is faster to put down and stronger.

## Plywood Subflooring

Normally, a plywood subfloor is put down in what is called a double system. First, the joists are covered with the cheapest ½-inch plywood you can find. Then, to cover the irregularities and to make the flooring smooth, the plywood is covered with either ⅛-inch hardboard or ½-inch particleboard. Use hardboard when the finished floor will be tile or linoleum; use particleboard when the floor will be carpeted. When using particleboard, which can also be covered with tile or linoleum, you should run it throughout the house so floors that are not carpeted will be the same height as floors that are carpeted.

Plywood panels are cut very precisely in 4-by-8 foot panels, and it is better to rely on their straightness than on your joist alignment. So the first step is to snap a chalkline across the joists 8 feet in from one side. Align the panels along this line, not on the rim joist. Lay the panels with the long way parallel to the joists for maximum strength. It's also a good practice to coat the joist tops with glue before putting down the plywood. This helps reduce the chances of squeaks.

Start the first course down the longest side of the building, using a full 4-by-8 sheet. Nail this in place with 8d nails spaced 6 inches apart along the edges and 10 to 12 inches apart along the joists. Then put the remaining panels in place with a 1⁄16-inch space between the ends. Use a couple of 4d nails for the spacing. (Double these spacing measurements in wet or humid areas.) Mark and cut the last panel to fit flush with the end joist.

Start the second course with a half panel (4 by 4 feet) so you will not be nailing the ends of two adjoining plywood sheets on the same joists. Space the sides of the plywood ⅛ inch from each other. This spacing allows for expansion and prevents floor buckling.

If, because of an odd building dimension, you know you're going to come out an inch or so short of the end, you can nail a 2-by-4 *scab*, or block, on the joist and cover it with a piece of scrap. Alternatively, you can increase the spaces between the ends of the panels slightly to make up this difference. These gaps will be covered by the hardboard or particleboard.

Drive the nails through the panels at a slight angle to help prevent them from working loose. All nails should hit the joists squarely. If you miss a section, squeaks may develop there. It doesn't take much time to snap chalklines for accurate nailing. This is especially important when you have to jog in the middle because of lapped joists.

### Hardboard or Particleboard Covering

The rough plywood is normally covered with ⅛-inch untempered hardboard for tile floors or ½-inch particleboard for carpeted floors to provide a smooth surface. If you try to cover the plywood with resilient tile or carpeting, breaks will soon occur over each knothole or deep irregularity.

Note that the hardboard or particleboard should not be put down until the roof and siding are completed. This is because the plywood will weather rainstorms but the hardboard will not. Also, water seeping between the plywood and hardboard or particleboard will set up conditions for rot.

Put the hardboard down in the same way you put down the plywood, but start the first course with a half sheet so that you do not end up with the same spacing arrangement. Do not space between the ends or the sides, and use 6d nails spaced every 6 inches. The stud walls should be in place when you put down the hardboard, so you will have to do a little more trimming along one end and one side.

**Plywood Subfloor**

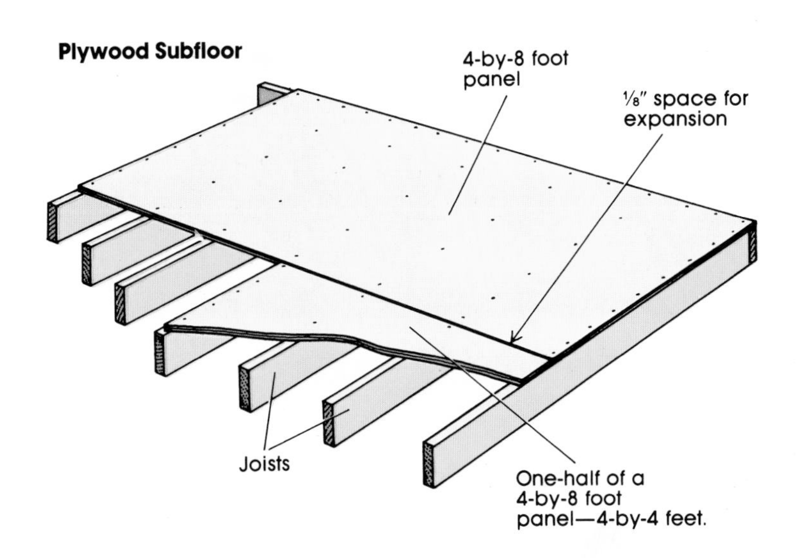


**Nailing Pattern**

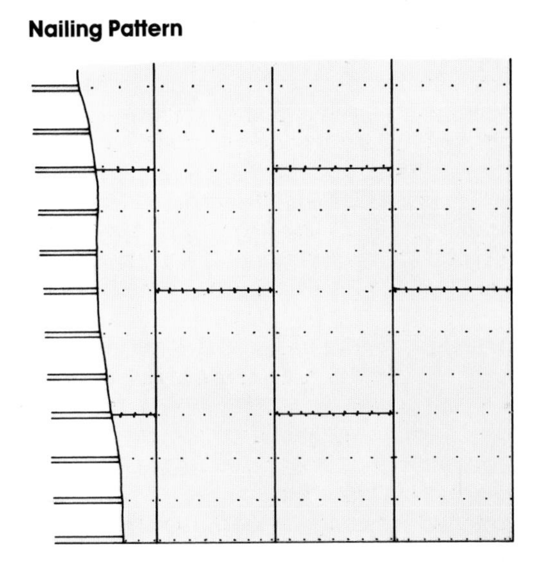

# FRAMING THE WALLS

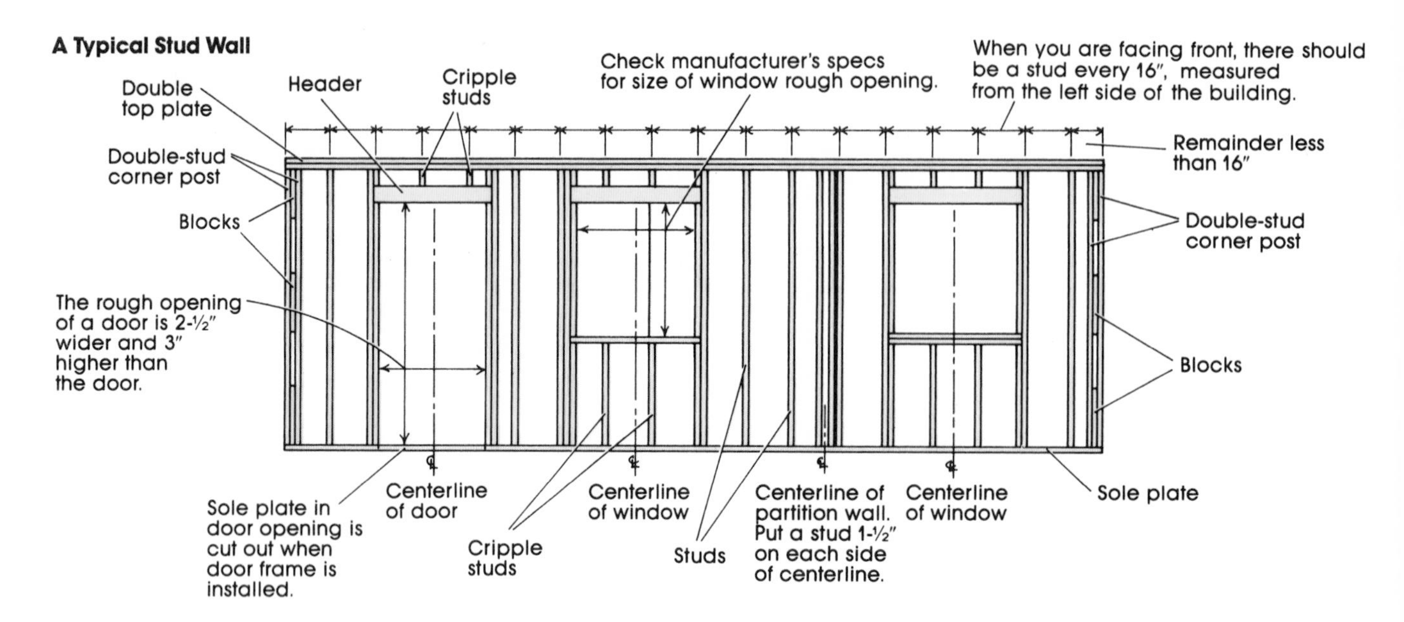


There are two methods of framing houses. One style, called balloon framing, has largely been replaced by a more efficient style called the western platform method, which will be covered here.

Before reading further, study the illustration and memorize all the terminology for stud wall construction.

## Constructing the Walls

The first thing to recognize is that your platform may not be perfectly square. But the stud walls *must* be square or the plywood siding will not fit right and your door and window openings will be crooked. Here's how you make the final corrections and at the same time begin laying out the wall.

For the top and bottom plates in a wall, use straight 2 by 4s at least 12 feet long; 16 feet is even better if you have someone to help you raise the wall once it is built. One person can raise a 12-foot wall section without anyone else's help. But the longer the wall, the easier it is to keep straight once erected.

Start the layout procedure by placing long, straight 2 by 4s flat on the subfloor down the longest side of the house. Place the outer edges of the 2 by 4s flush with the edge of the subfloor platform and the ends flush with the ends of the platform. These 2-by-4 lengths will form the sole plate. Wherever two pieces are butted together, they must fall on a 16-inch mark. A stud will bridge that joint and support both pieces. You cannot have a joint in the sole or top plate unsupported by a stud. If all but the last length of 2 by 4s are in multiples of 4 feet, the joints will always come out over a stud. The last length should be at least 4 feet long. If necessary, cut the next to last board shorter (at a 16-inch interval) so the last piece will still be at least 4 feet long. Lay out the sole plates down the other side in the same manner and then fill in both ends.

If your subfloor platform is square, you will have no problem. If not, adjust the sole plates until they are square. When the measurements of each diagonal are equal, the sole plates are square. If the plates must come in or hang out over the platform by a ¼ inch, it is not good but is acceptable. The siding will extend beyond that and hide the discrepancy.

When the sole plates are aligned, tack them in place with 8d nails. Don't drive the nails in because you will soon remove them. Next, with pencil or marking crayon, outline on the subfloor the interior edges of all the sole plates. When the wall is raised, position the sole plates on these lines again.

Now cut the same number of matching pieces. These will make up the top plate. Place them next to the sole plates, but start from the opposite end so you will not have top and sole plate joints over the same stud, which would make a weak connection. *A butt joint on the sole plate must be at least 4 feet away from a butt joint on the top plate.*

The next step, with the sole and top plates still side by side, is to mark all the door and window openings on the plates. (For details on constructing door and window frames, see pages 91–92.) Laying out the door and window openings first will make it clear to you where studs can be eliminated.

Interior walls are laid out in the same manner as exterior walls.

### Marking Plates for Studs

Clear the subfloor except for the sole and top plates for the longest wall. Stack the lumber carefully so you can put it back correctly when you start the next wall. Always build the two side walls first and then the end walls.

With the sole and top plate aligned down one side, measure from one end of the sole plate and mark at 16 inches. The first stud will be centered on that mark. However, since the mark will be covered by the stud when you start nailing, here's a trick for faster and more accurate alignment: instead of marking exactly on the 16-inch intervals, come back ¾ inch and mark. That is where the edge of the stud will go.

Put your combination square on this mark and draw a line simultaneously across both plates. If you are working from left to right, the line represents the left edge of the stud. Put an X to the right of that line on both plates to remind you which side of the line the stud goes on.

### Putting the Wall Together

The wall is built on the subfloor platform, then raised and braced in position. The process is essentially the same whether or not this section of wall has doors and windows.

The first step is to remove the nails from the sole plates. Move the top plate away from the sole plate a little more than the length of the studs and stand all plates on edge, with the marks facing each other.

Studs are all a standard 92¼ inches long. Precut studs come in this length; if you are using 8-foot 2 by 4s, first cut them all to length. The studs are 92¼ inches long so that when you add the width of the sole plate, the top plate, and the cap plate (often called the double top plate), the overall height of the wall is 96¾ inches. This allows you ⅜-inch clearance top and bottom when putting up the interior wall covering of sheetrock or paneling.

After the plates are spaced apart, the next step is to build the corner posts. Both side walls must have these posts at each end. The end walls do not have corner posts. The single stud at both ends of each end wall is nailed to the corner posts when the end wall is raised in place.

Make the corner posts by sandwiching 3 pieces of 2-by-4 scrap, each about 12 inches long, between two studs. Lay out the corner posts and studs on edge and butt them against the markings on the sole plate. Stand with one foot on the stud and one foot on the sole plate to keep them from moving, and nail two 16d nails through the sole plate into the bottom of each stud. Do the same when you nail on the top plate. You now have a basic stud wall. It's obviously not square, lying there on the subfloor, but don't worry. You'll do that after you stand it up.

**Corner Post Construction**

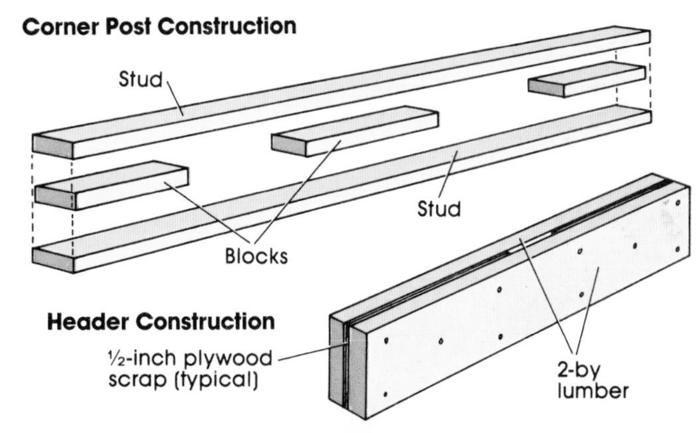


### Framing Window Openings

The tops of all windows and doors should be the same height. Standard house construction calls for that height to be 80 inches from the top of the sole plate to the bottom of the header. Therefore, cut all trimmer studs, which support the headers, 80 inches long. The width of the doors and the width and height of the windows are determined by your plans. Choosing these sizes is a matter of personal taste.

The rough window opening should be ½ inch larger all around than the actual window. This allows you room to maneuver the window slightly before nailing it to ensure that it is square and will not bind when it is opened. Trim will cover the small spaces between the window, trimmers, and headers.

A header must be placed above this opening for structural support. Headers can be fabricated from pieces of 2-by or can be precut to length at the lumberyard from a 4 by 12. Study the illustration to determine the length of your headers. Note that the trimmer studs on each side

**Marking Plates for Studs**

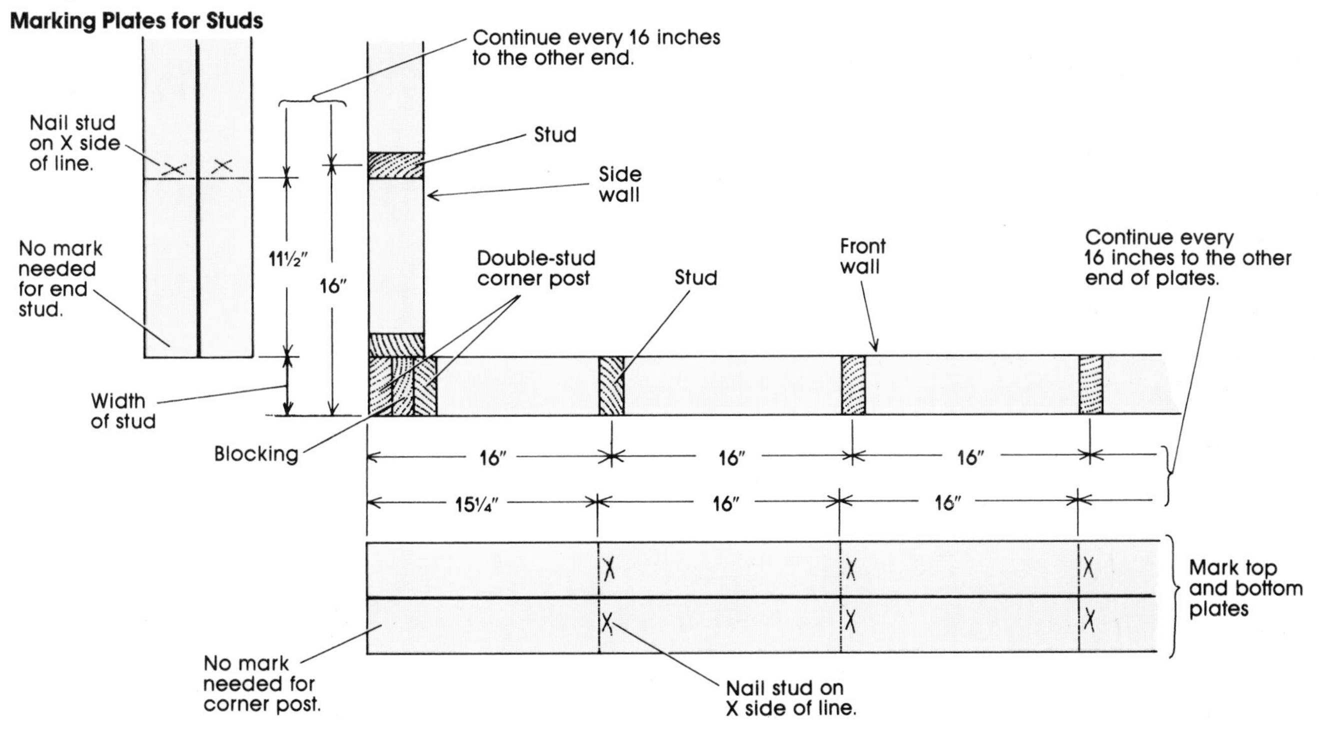

give the rough opening for the window and support the header. Therefore, the header must be 3 inches longer than the rough opening to accommodate the two trimmer studs.

If you make all headers from 4 by 12s, your work is simplified. If you fabricate your own, use this guide to determine the width of the header boards:

Maximum Span (ft) 3½ 5 6½ 8

Header Size (in.) 2 by 6 2 by 8 2 by 10 2 by 12

Fabricated headers must have a ½-inch piece of plywood sandwiched between them to make them the same width as a stud.

When you lay out the window openings on your sole and top plates, study the illustration carefully to note the locations of the opening studs and the trimmers. Put the opening stud in place, or use an existing stud if possible, then cut the trimmer studs 80 inches long and nail them to the opening studs. Nail the header in place through the opening studs.

If you have fabricated your own header, there will most likely be a space of anywhere from 2 to 6 inches between the top of the header and the top plate. This space must be supported with short 2-by-4 blocks called *cripples*. Put them where studs would normally go.

The bottom of the window opening consists of a 2-by-4 sill that fits between the trimmers. Some codes call for double sills.

Cripples are also needed below the window to support the sill. Note that the top and bottom cripples should be in line with each other and should be placed every 16 inches o.c., just as studs would be placed, so that you can nail on siding and interior panels correctly.

### Framing Door Openings

Framing a door opening is very similar to framing a window opening, but you have to take into consideration the thickness of the doorjamb (the framing around the door itself). Whether you buy a prehung door, which comes complete with the jamb, or make your own jamb and door, the same process is involved in framing the opening.

Doors are a standard 80 inches high, so the door opening is made 80 inches high when it is being framed. After all the walls are up and you are ready to install the door, cut out the sole plate in the opening. This additional 1½ inches makes room for the top jamb and the door sill.

Many plans give measurements from the end of the house to the center of the door. Mark that point with a large C on the sole plate. You should be able to adjust the opening a few inches to the left or right so you can use an existing stud for the rough opening stud.

Divide the width of the door in half and measure that distance each way from the center mark. For example, for a 36-inch exterior door, measure 18 inches to the left and 18 inches to the right of center. This gives you the actual door opening.

Next, add the thickness of the door jamb on each side. Standard exterior door jambs are 1½ inches thick, so add 1½ inches on each side.

If you could build a perfectly straight and plumb house, you would put the inside edge of the trimmer right at that point and just slip in the jamb and door. But no carpenter counts on being perfect. Instead, add an additional ¼ inch to each side and put the inside edge of the trimmer there. Later you will fill this space with shims made from shingles to square up the door jamb (see pages 102–103).

Cut the trimmers 80 inches long and install one of them next to the rough opening stud. Install the other trimmer at the marked point and then install the header, as explained in the section on framing windows.

**Framing a Window** **Framing a Door**

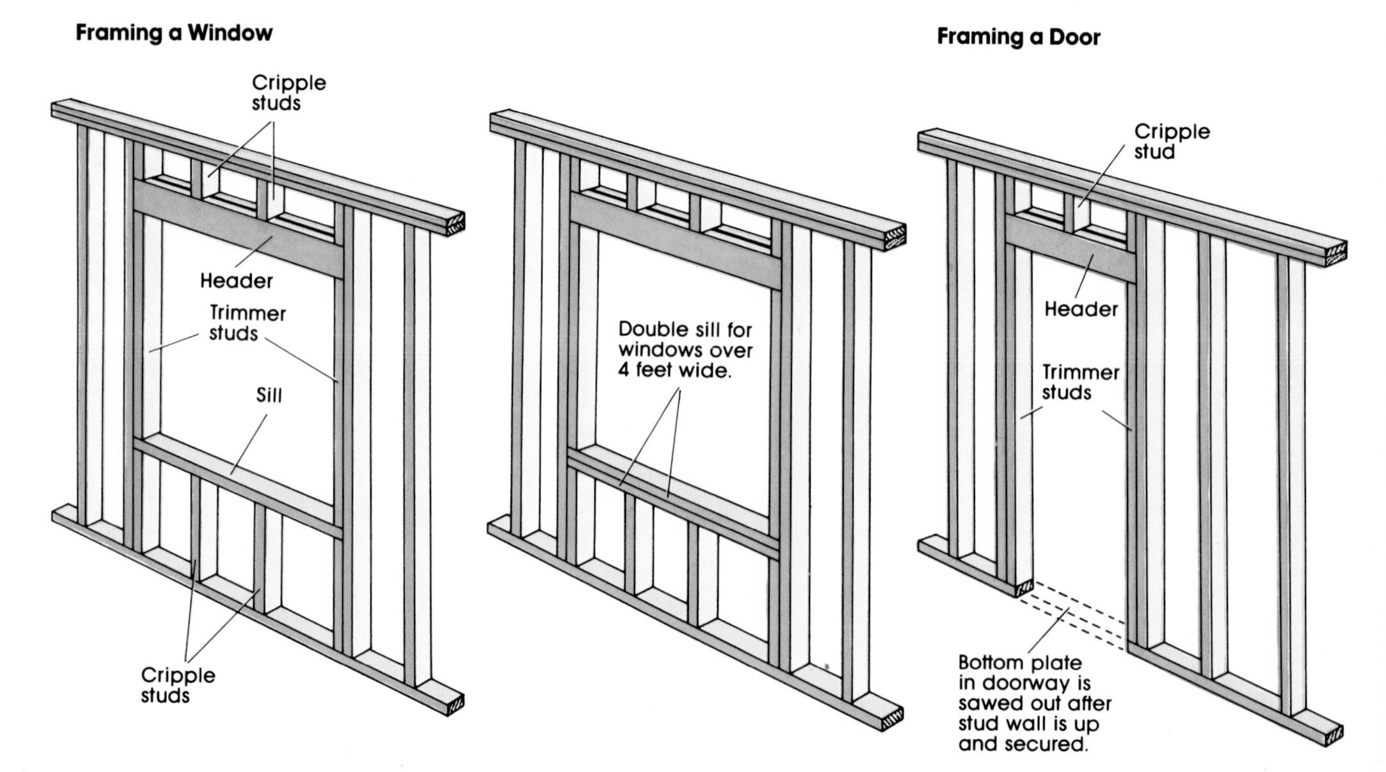

Let-in Bracing

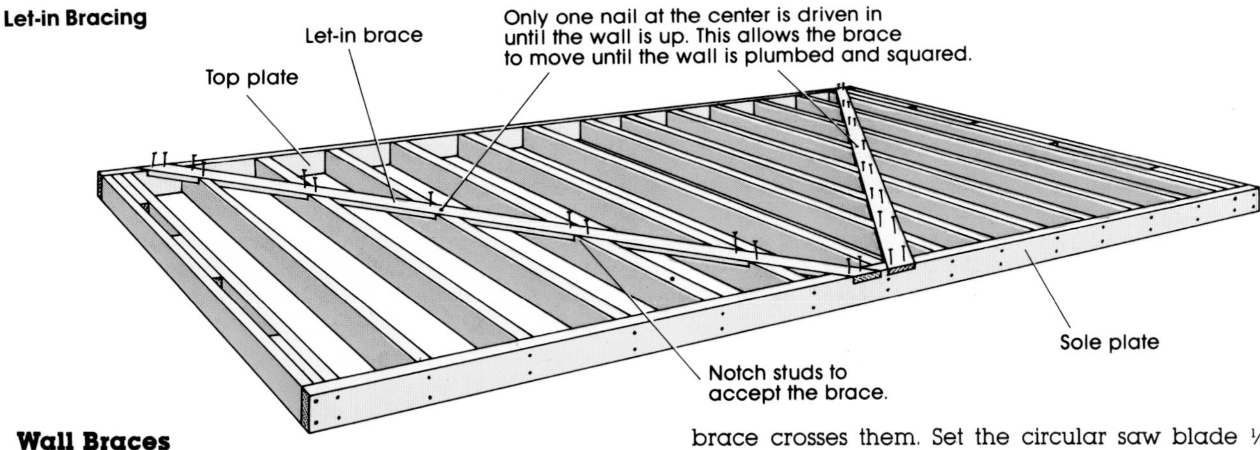


## Wall Braces

Buildings with plywood siding normally do not require bracing, but all others do. The two most widely used types are let-in bracing and metal strap bracing.

**Let-in bracing** is made from 1-by-4 stock and runs from the top outside corners of the wall to the bottom center to form a V shape. It is set into notched studs and is prepared while the stud wall is still lying on the platform.

These braces may seem to involve a lot of detail work, but it takes just a few minutes to make each one, and they add a great deal of strength to the wall.

While the wall is still lying flat on the subfloor, push it on one corner or the other until it is close to being square. Lay the 1 by 4 on the wall with one end at a top corner and the other end as far out on the sole plate as possible without running into any door or window openings. Mark the underside of the 1 by 4 where it overhangs the sole and top plates to get the angle at which the plates cross. Mark both sides of the studs and plates at each point the brace crosses them. Set the circular saw blade ⅛ inch deeper than the thickness of the 1 by 4 and cut every line you marked on the studs and plates. Now make a series of closely spaced cuts between these lines. Knock out the wood with your hammer and chisel off any stubborn chips. Trim the ends of the 1 by 4 and put the brace in place. Hold it there with a single nail in the middle. Don't nail the tops and bottoms until you stand all the walls up and plumb and square them. Then put two 8d nails wherever the brace crosses a stud or plate.

**Metal strap bracing** is commonly available in 10- and 12-foot lengths and is nailed to the outside of the stud walls after they are up, squared, and plumb. The braces are thin enough not to obstruct the exterior wall sheathing.

Metal strap braces come with holes for 8d nails drilled every 2 inches. To attach them, first make sure the walls are squared and plumb, then nail one end of the brace to the top plate, nail it to every stud it crosses, and nail the other end to the sole plate. Metal strap bracing must always be put up in crossed pairs, as illustrated.

Metal Strap Bracing

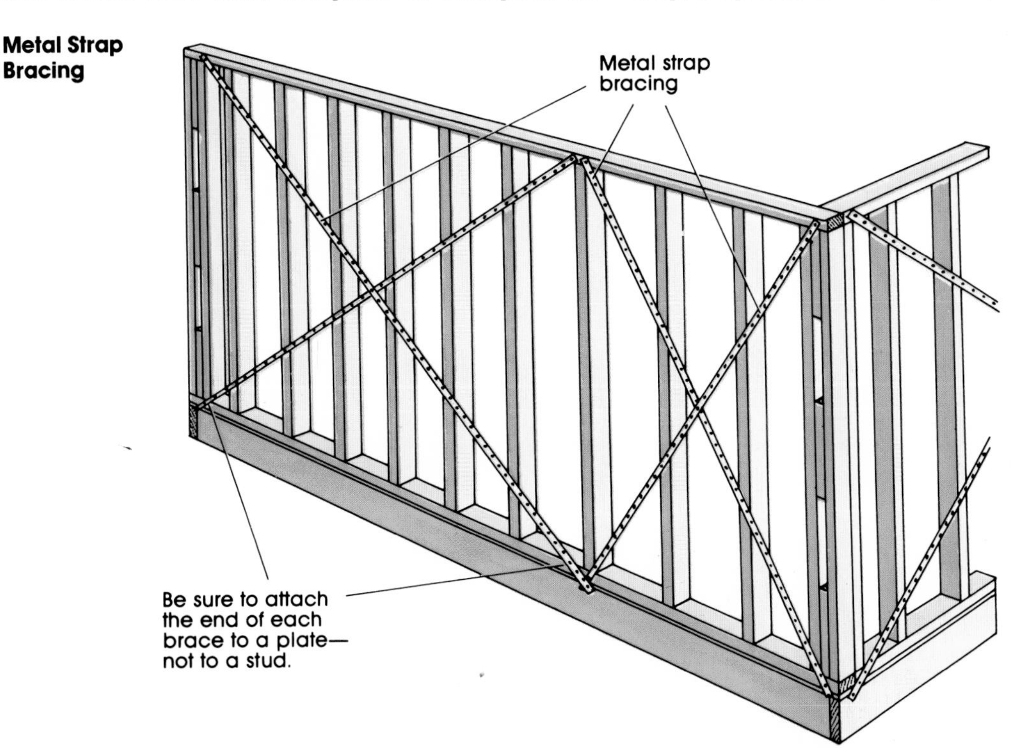

### Raising, Plumbing, and Squaring Walls

The two long side walls are built and erected first, and then the two end walls. Two people can raise walls that are 16 to 24 feet long. Normally the wall will not slip off the platform, but if you are worried, tack a couple of pieces of 1 by 4 to the rim joist to catch the sole plate.

Raise the wall smoothly and evenly, being particularly careful that it doesn't buckle at the butt joints. When it's up, one person can steady it while the other braces it.

The first brace, a 1 by 4, goes from the top end of the wall to the outside of the end joist. Put the intervening braces about every 12 feet. Tack some 2-by-4 scrap to the subfloor so you can nail the end of the brace into it. (The braces will be removed after the roof is up so you can put on the siding.)

First nail the braces with one nail high up on the studs. Then tap the sole plate with your hammer until it is perfectly in line with your marks on the subfloor. Put a few nails through the sole plate into the floor to hold it. Later go back and put two nails through the plate between each joist.

With the plate secured, one person uses the level to plumb the wall while the other secures the braces.

Put the end walls in place after the side walls are up, plumbed, and braced. Align the sole plates, make sure everything is flush on the outside, and then nail the end studs to the corner posts of the side wall.

The final step is to make sure your walls are square. Use the plumb board to check the ends. One person on a ladder at the end of the wall with the plumb board can push or pull the wall until it is square while the other person tacks the braces down. Don't drive the nails completely until all walls have been squared. Measure the diagonals from the top corners of the walls to check your work. If the walls are square, finish nailing the bracing.

Long walls have a tendency to bow in or out in the middle. Before you nail on the cap plate, sight down the wall to make sure it is straight. Mason's twine pulled from end to end, with a little block of wood to hold it out from the top plate, will give you a more accurate sighting. Make any necessary adjustments with the center bracing to straighten the wall.

The installation of the cap plates is the final step in the construction of the stud wall. The cap plates on the end walls overlap the side walls, tying the walls together at the top. Cap plates on any interior walls must tie onto an exterior wall. Any butted ends in the cap plate must be at least 4 feet from butted ends in the top plate. Select the straightest possible 2 by 4s for your cap plates.

Interior walls are laid out and constructed in the same manner as exterior walls. They are designed not only to divide the house into rooms but also to support ceiling joists if the house is too wide for a single span. A wall running parallel with a ceiling joist, and carrying the weight of that joist, should be supported underneath with a double floor joist.

If you want an open section in the house interior, but still need to support ceiling joists, then you must span that opening with a properly sized beam (see span chart on pages 108–110).

When all the walls are completed, you are ready to install the ceiling joists.

**Bracing a Stud Wall** **Plumb Board**

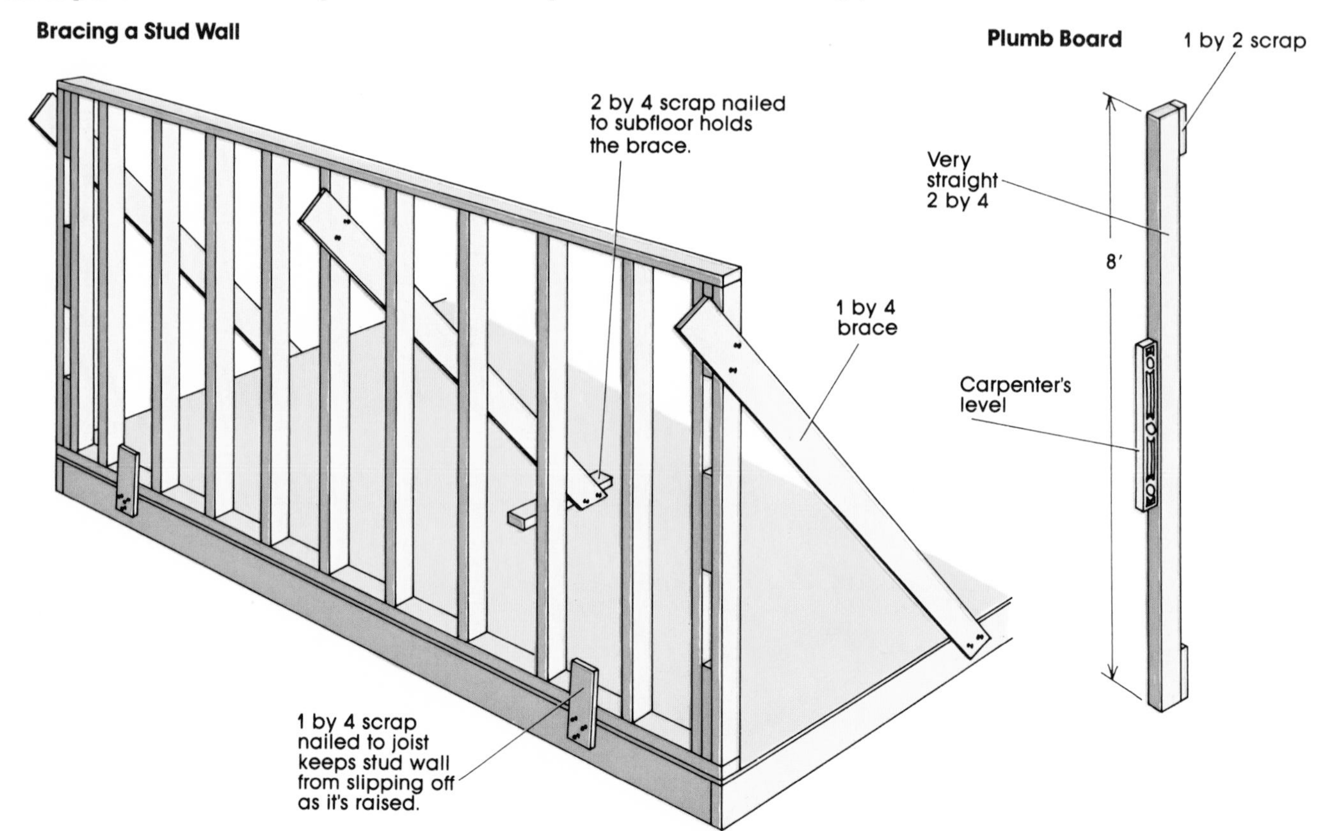

# CEILING JOISTS

Ceiling joists tie the house together to resist the outward pressure of the rafters and provide nailing space for the ceiling. Because ceiling joists fit alongside the rafters, you need to mark the rafter locations first. Therefore, before proceeding with ceiling joists, read the section on rafters (pages 96–98) so you can plan and mark locations accurately. Rafters are commonly spaced 2 feet o.c., but in areas with heavy snowloads you may be required to space them 16 inches o.c.

Mark the rafter locations on top of the cap plate just as you did for studs. Starting from the same end of each side wall, measure off 16 inches, then back up ¾ of an inch and use your square to make a straight line. Put an X to the right of the line. The rafter will be placed over the X. Place the ceiling joists on the side of the line opposite the X.

The ends of the ceiling joists should be flush with the outer edge of the building. The part of each joist that extends above the rafter is trimmed off after rafters and joists are all in place.

Position each ceiling joist correctly and toenail both ends to the cap plate. After the rafters are in place, nail the rafters to the joists.

Ceiling joists are normally made from 2 by 6s, which will not span the entire width of standard houses. They must be supported in the middle by interior walls or beams. When joists are butted together on an interior wall, they must be tied together with 2-foot strips of ⅜-inch plywood on both sides.

Alternatively, you can place all the joists down one side of the building on one side of the rafters, and all the joists down the other side on the opposite side of the rafters. Fill the gap in the center where each pair of joists meet with a 2-by-6 block. Toenail the joists to the top plate of the bearing wall.

End joists cannot be placed directly against the end rafter because the gable end studs extend inwards (see page 98 for details on gable end studs). Nail the end joists to the gable end studs and fill the gaps on each end with a piece of blocking.

## Collar Beams

If you want open roof space overhead instead of a ceiling, you need to plan ahead so that the underside of the roof will have a finished appearance. You will also need to plan for collar beams, which prevent the weight of the rafters and roof from spreading the walls outward.

Collar beams can be made from 2 x 4s and are nailed across both sides of each pair of rafters, as shown in the illustration. Measure down from the ridgeboard about one-third the rafter's length and place the collar beams there.

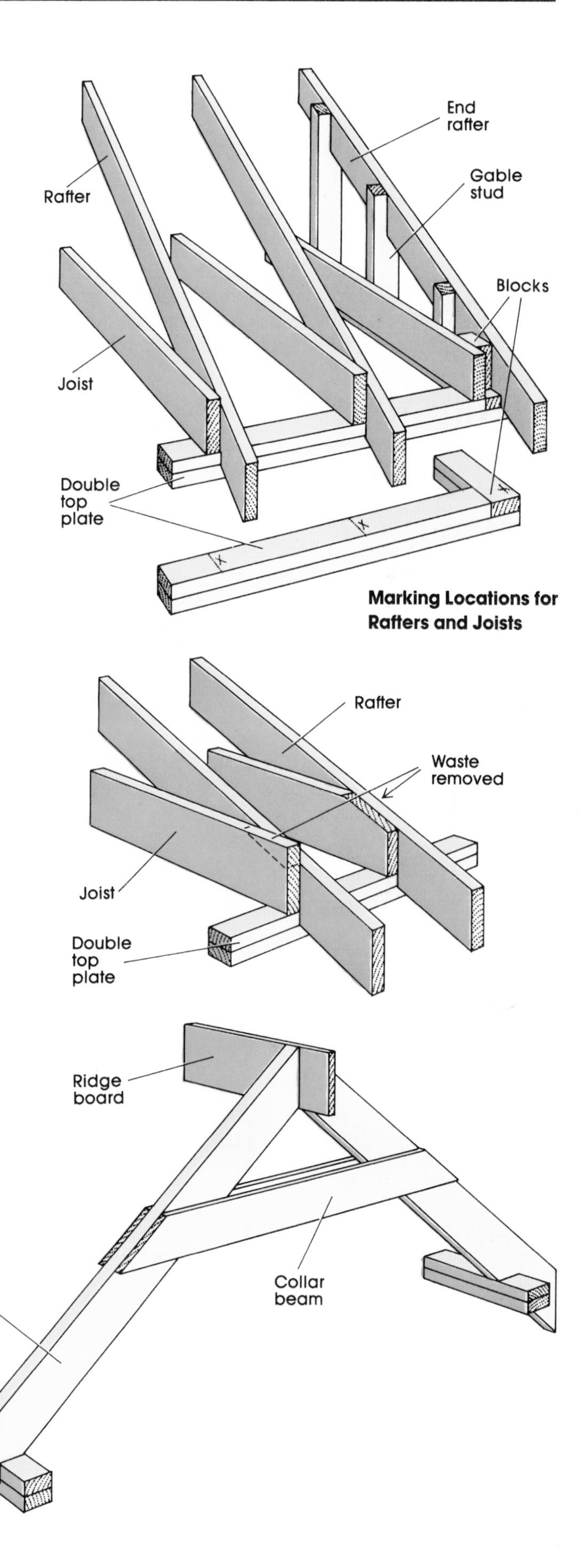


**Marking Locations for Rafters and Joists**

# FRAMING THE ROOF

There are many types of roofs. This book deals only with the standard gable roof, which consists of common rafters, a ridge board, gable end studs, and ceiling joists or collar beams. It does not deal with roofs that involve the more complex hip, valley, and jack rafters.

The design of the gable roof is based on the formation of a right triangle. The length of the cap plate on the stud wall from the edge of the building to the center of the end wall forms the bottom of the triangle. The height of the roof represents the vertical part of the triangle, and the rafters form the hypotenuse or third side of the triangle. The tables on your framing square will show you how long the rafters must be.

## Roof Pitches

All roofs must have a slope, or *pitch,* to shed moisture. Pitches range from nearly flat to a 45 degree angle, such as for an A-frame roof. You must decide what pitch you want.

You should commit the following terms to memory: the *span* is the width of the house; the *run* is half the width; and the *rise* is the distance from the top of the cap plate to the top of the roof.

There are four principal pitches for roofs: ⅙, ¼, ⅓, and ½. A pitch of ¼, for example, means that the rise is one-quarter the length of the span. A look at the following illustration will show you that many pitches are possible besides the four that are most commonly used.

In addition to the fraction terminology just described, carpenters also refer to 4 in 12 pitch, 6 in 12 pitch, and so on. A 6 in 12 pitch, for example, means that the roof rises 6 inches for every foot of run. This terminology correlates with the fraction terminology described above as follows:

⅙ = 4 in 12
¼ = 6 in 12
⅓ = 8 in 12
½ = 12 in 12

When you have selected the pitch you want, you need to calculate how high the rise will be and how long the rafters must be. The following formula will give you the rise:

Pitch × Span = Total rise

For example, if a building is 24 feet wide and requires a ¼ pitch, the rise should be

¼ × 24 feet = 6 feet

The next step is to determine the length of the rafters. First, study the illustration on this page and familiarize yourself with the terms used to describe a common rafter.

In the example given above, the building is 24 feet wide and will have a ¼ (6 in 12) pitch roof. To determine the rafter length, look under the number 6 on the face of your framing square blade. You will see the figures "13 42." This means that for every 12 inches of run (half the span) in the building, 13 42/100 inches of rafter length are needed. In our example the run is half the 24-foot span, or 12 feet.

To determine the rafter length, multiply the length given in the table on your framing square (13.42 inches in this example) by the number of feet in the run. In this case the rafter length will be

12 × 13.42 inches = 161.04 inches

You can disregard the .04; it's too small a fraction to work with for rafters.

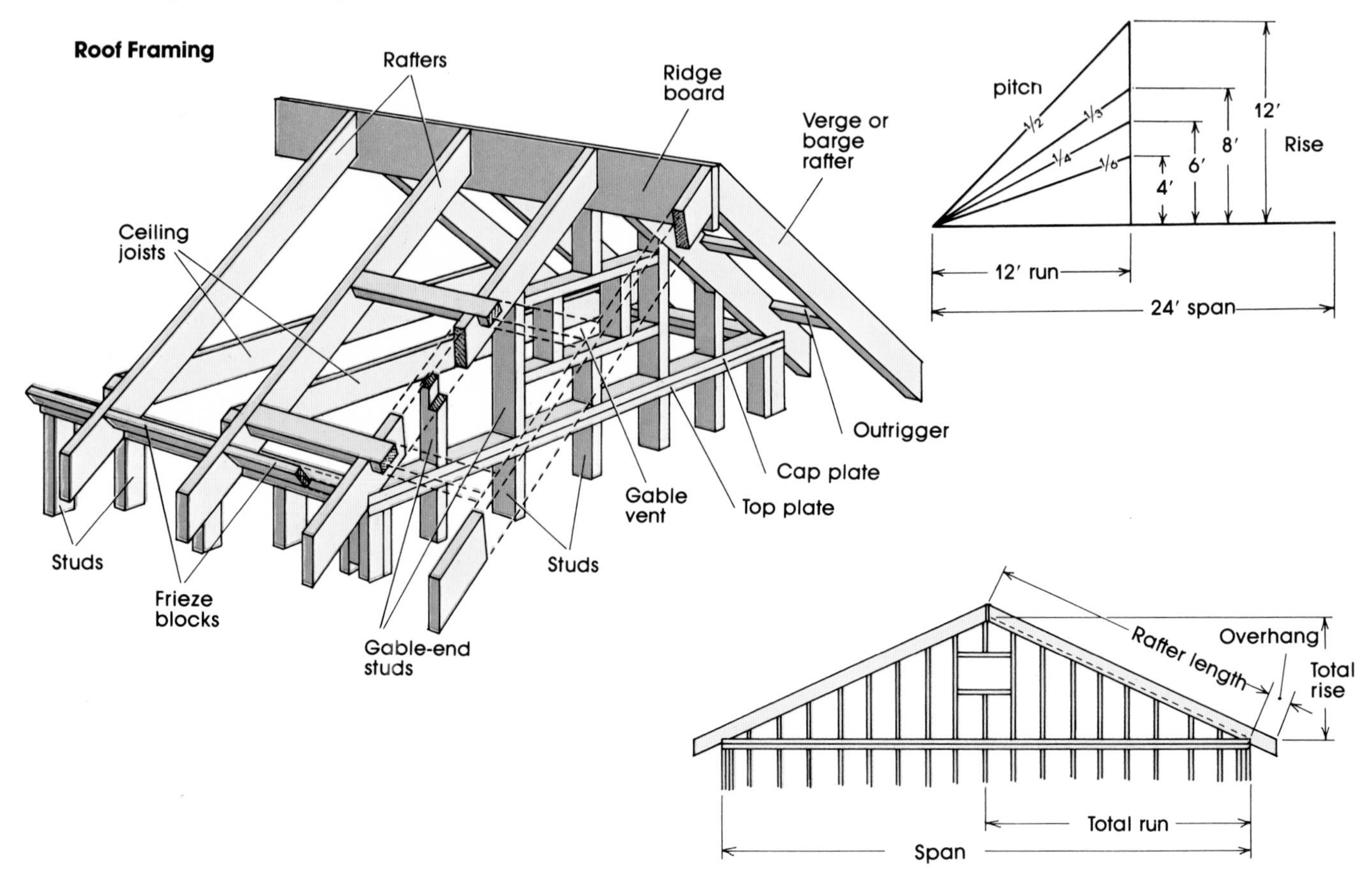

To convert the inches to feet, divide by 12 (a pocket calculator will make life much easier for you):

161 inches ÷ 12 = 13 5/12 feet

or 13 feet 5 inches.

When you have completed all your rafter calculations, you are ready to start framing the roof. It takes three people to frame a roof unless you have a lot of experience.

## Framing Procedure

**1.** Mark the top plates of the walls for rafter locations if you have not already done so (see page 95 for details on marking ceiling joist and rafter locations).

**2.** Cut the ridgeboard, allowing for the amount of gable overhang specified by your plans. (An 18-inch overhang is standard). The ridgeboard is always one size wider than the rafters: for example, 2-by-6 rafters take a 1-by-8 ridgeboard. If your house is too long for a single ridgeboard, cut it so that each section butts together where a pair of rafters meet and not in a span between rafters.

**3.** Lay out the rafter locations on the ridgeboard, again allowing for the amount of gable overhang specified by your plans.

**4.** Cut the first pair of rafters. Remember to trim half the thickness of the ridgeboard from the ridge end of the rafter (see box on this page). Put this first pair in place at one

### Cutting Rafters

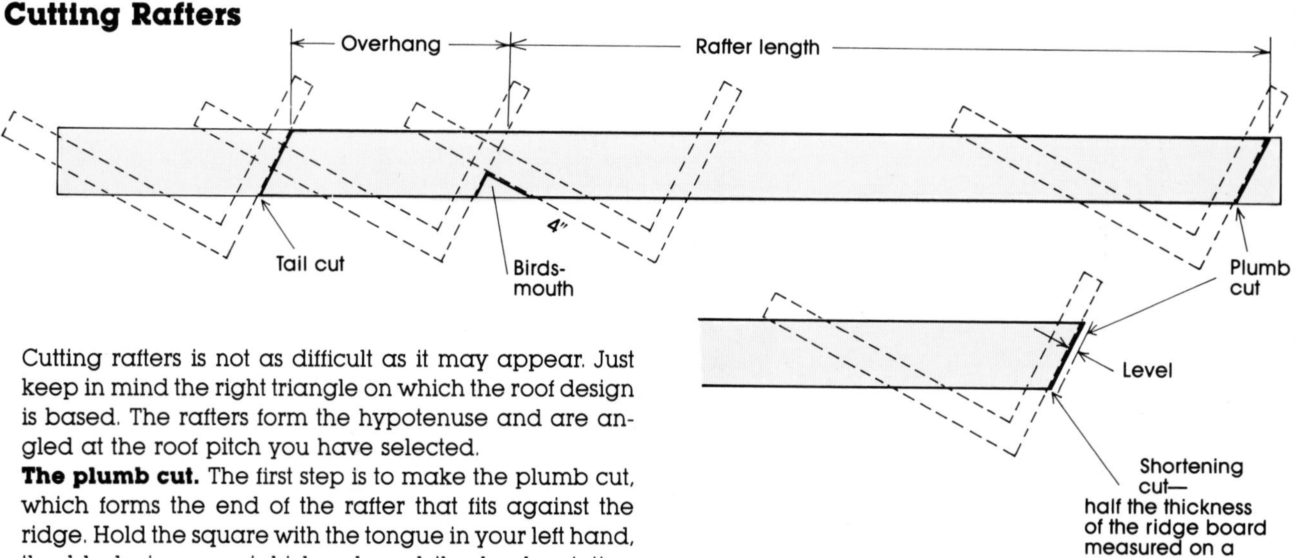


Cutting rafters is not as difficult as it may appear. Just keep in mind the right triangle on which the roof design is based. The rafters form the hypotenuse and are angled at the roof pitch you have selected.

**The plumb cut.** The first step is to make the plumb cut, which forms the end of the rafter that fits against the ridge. Hold the square with the tongue in your left hand, the blade in your right hand, and the heel pointing away from you. For the 6 in 12 pitch called for in the example on this page, place the 6-inch mark on the outside edge of the tongue on the edge of the board nearest you. Next, put the 12-inch mark on the outside edge of the blade on the board's edge. Draw a pencil line along the outside edge of the tongue. That is the required angle for a 6 in 12 cut. (As another example, suppose you had chosen a 4 in 12 pitch. You would first position the 4-inch mark on the tongue, then the 12-inch mark on the blade. The 12-inch mark on the blade remains constant; the pitch adjustment is made by moving the tongue).

**The birdsmouth.** After making the plumb cut, measure off 161 inches (to continue the example) from the tip of the rafter. Place the 6-inch mark of the tongue on this point, and align the 12-inch mark on the blade as you did previously. Again, draw a pencil line along the outside of the tongue.

Now, still holding your 6- and 12-inch marks, slide the square to your left, toward the plumb cut. Position it so the distance from the outer edge of the board to the line you have just drawn is 4 inches. Draw a pencil line along the outside edge of the blade. You will cut out the section formed by the two lines you have just drawn. This birdsmouth will allow the rafter to rest flush on the top plate of the stud wall.

**The tail cut.** To make the tail cut, slide the square down toward the rafter end for as much overhang as you want (18 inches from the birdsmouth is common). Mark along the outside edge of the tongue. This angle should match that of the plumb cut. Eventually, you will nail a fascia board or a rain gutter to the exposed rafter ends.

**The shortening cut.** This is a simple step, but it must not be overlooked. Rafter lengths are first measured as if they butted against each other at the peak of the roof. But roofs are normally constructed with a 1-by ridgeboard that makes it easier to put up the rafters and gives them more stability.

For the rafters to fit properly, half the thickness of the ridgeboard must be trimmed off each rafter at the plumb cut. If using 1-by stock, which is 3/4 inch thick, trim off 3/8 inch from the rafter end. If using 2-by stock, which is 1½ inch thick, trim off 3/4 inch. The cut must be absolutely parallel to your initial cut. Lay it out with your square again.

In addition to the rafters that rest on the top plate of the wall, you also need four barge rafters, which are the same as the others except that they don't have the birdsmouth cut. These are put up last and fit at the ends of the gable overhang. They are supported by outriggers (see page 98).

end to make sure you have an accurate fit. If everything is correct, use one as a pattern and cut all the remaining rafters.

**5.** Now you are ready to start erecting the rafters. The first few pairs are the most difficult. One method is to put up a notched 1 by 4 at each end to hold the ridgeboard while you nail on the rafters. If you have three people, however, one person holds the two end rafters and the ridgeboard in place, another holds the other end of the ridgeboard level, and the third toenails the rafters to the cap plate. The pressure of the two rafters at the top will hold the ridgeboard in place until they are nailed to it.

**6.** With the first two pairs of rafters up, you should check three things immediately: that the ridgeboard is centered over the house, that it is level, and that the end rafters are plumb with the end of the house. When everything is correct, brace the ridgeboard with two 1 by 4s running from near the center down to the end wall studs.

**7.** Nail up the remaining rafters and nail the rafters to the ceiling joists where they meet on the side walls.

**8.** Nail in frieze blocks. These blocks are made from the same stock as the rafters and fit between each rafter to fill the hole between the rafters and the cap plate. They are not necessary if the underside of the rafter overhang is to be covered by a soffit, which is not described in this book. Each frieze block is angled slightly so that the top is flush with the top of the rafters and the bottom rests on the edge of the cap plate. Put frieze blocks in place at both ends of the roof and then snap a chalkline to align all the others.

**9.** Use a chalkline to check that the rafter tail cuts are all in line. Nail on the fascia board, which is the same length as the ridgeboard. Then nail the barge rafters to the ridgeboard, the fascia board, and the outriggers (see below for details on outriggers).

## Gable End Studs

When the roof members are up and braced, you need to fill the end walls formed by the rafters. The gable end studs fit directly above the wall studs and support the end rafters.

One way to determine the length of each gable end stud is to hold a short length of 2 by 4 in position behind the rafter—make sure it's plumb—and mark where the rafter crosses it. Cut it at that angle, then toenail the bottom of the stud to the cap plate and the bottom of the rafter.

It's almost as easy to cut a 1½-inch notch in the top of the gable end stud so that part of it supports the back of the rafter. Just make a 1½-inch-deep miter cut on the edge of the stud and then make a rip cut from the top of the stud down to the first cut.

## Outriggers

Outriggers provide the support required to extend the end gables beyond the end rafter by the standard 18 inches. They consist of lengths of 2 by 4 laid flat that extend from the first interior rafter across the end rafter and out to the barge rafter on the gable overhang. One end of the outrigger is butted to the first interior rafter, and the end rafter is notched 1½ inches so the top of the outrigger will be flush with it. The barge rafter is nailed to the protruding end of the outrigger.

Outriggers should be spaced 4 feet apart. Start measuring at the ridge.

## Vents

A vent on each end of the house, close up under the gable overhang, allows air to move freely through the attic and prevents the buildup of condensation. The vent is constructed by nailing two horizontal 2 by 4s between the gable end studs just left and right of center. Support the bottom of the vent opening with a 2 by 4 nailed in the center. Toenail another 2 by 4 in the center of the top crosspiece running up to the ridgeboard. Cover the opening with window screening to keep out insects, birds, and other wildlife.

**Frieze Blocks**

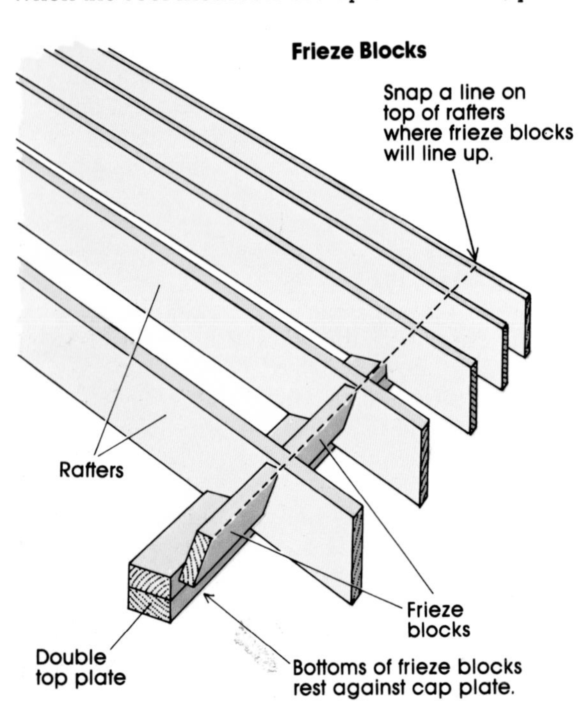


**Gable End Studs**

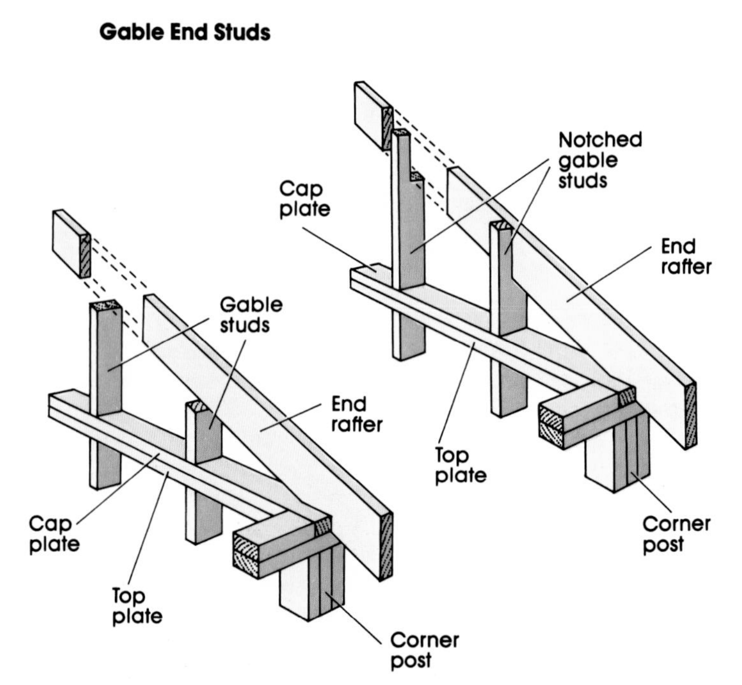

## Roof Trusses

Although a good carpenter should know how to cut and erect rafters, don't overlook the possibilities of roof trusses. These are ready-made units that include the ceiling joists and rafters. Because they are mass produced, they are nearly as cheap as building them all yourself. The framework in a truss will serve the same purpose as gable end studs for the end rafters.

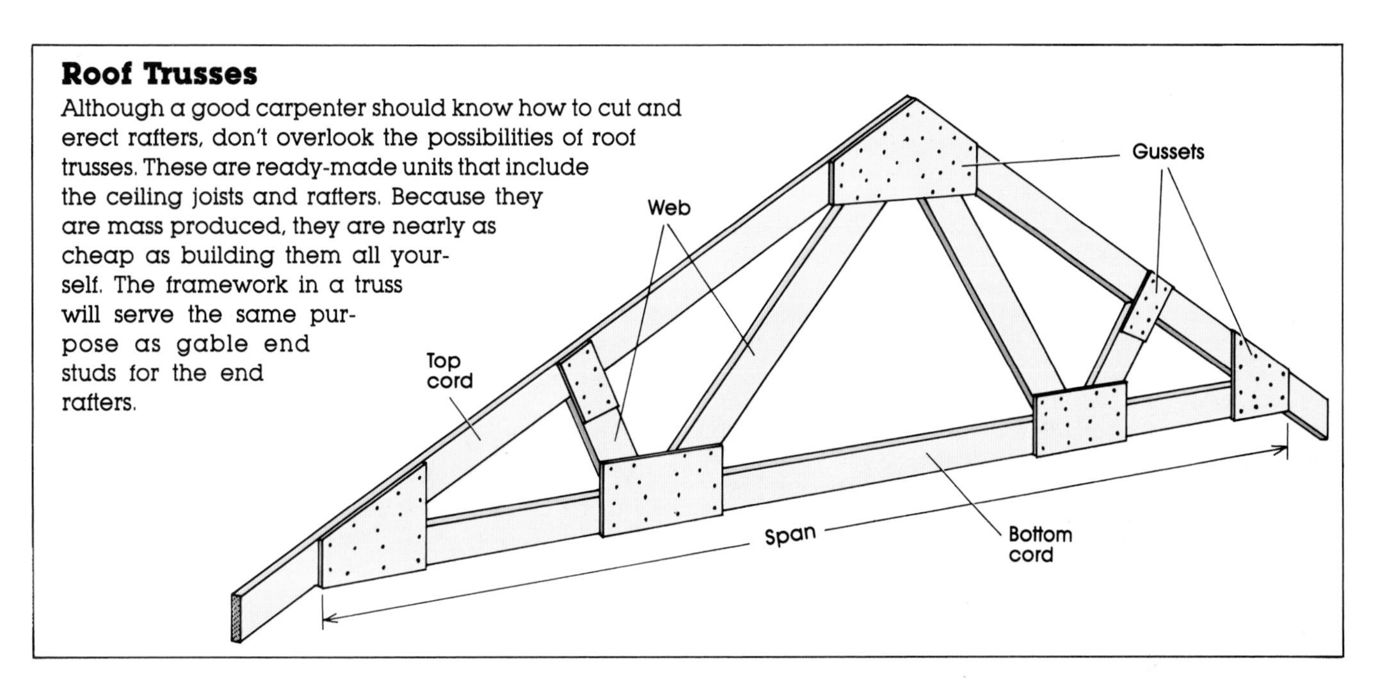


Roof Sheathing

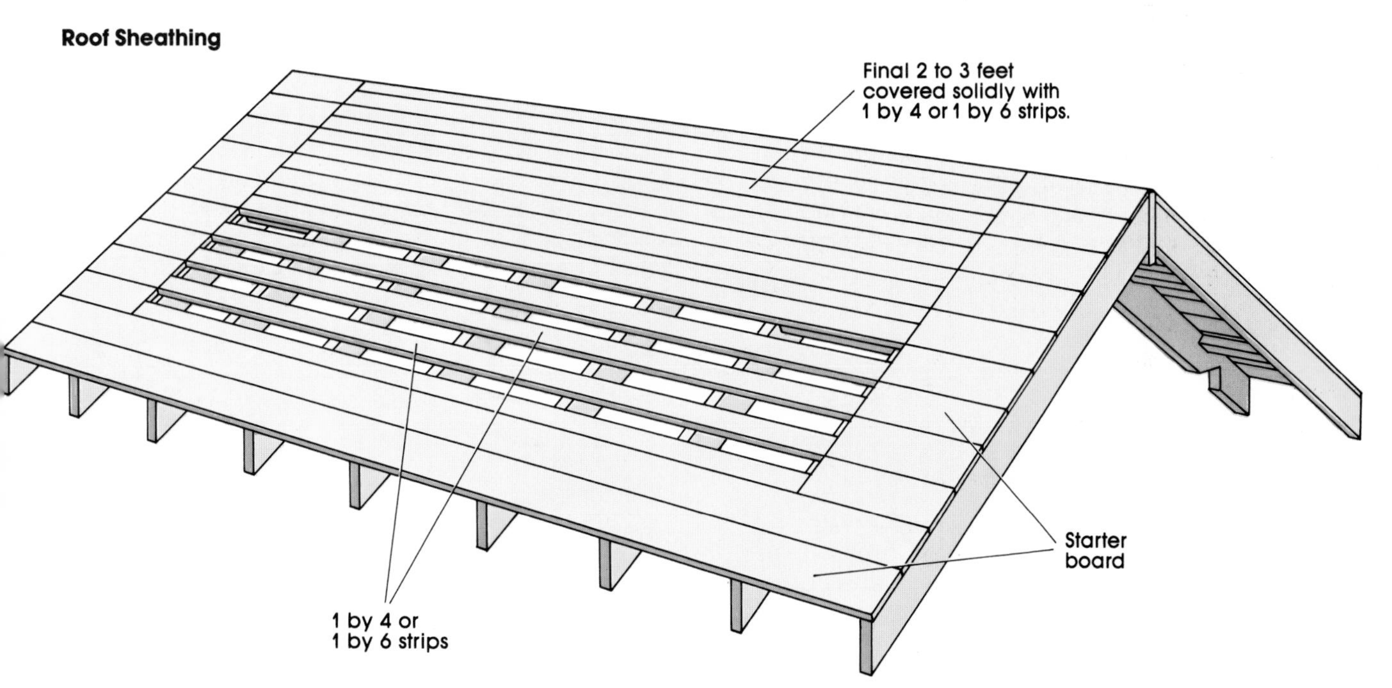


## Roof Sheathing

With the rafters and gable end walls up, the next step is sheathing the roof. Sheathing increases the rigidity of the roof and provides nailing support for the roofing material.

Two common choices for sheathing are solid plywood or spaced 1 by 4s, which are considerably cheaper. The steps involved in putting on 1-by-4 stripping for a shake roof, as was done for the playhouse, are as follows:

**1.** Put down starter board over the overhang space between the end rafters and the barge rafters. Starter board, which is 1-by-8 shiplap (see page 8), gives the exposed underside of the overhang a smooth and finished appearance. It should be cut so it is flush with the outer edge of the barge rafter and covers just half of the end rafter. This leaves nailing space for the 1 by 4 stripping.

**2.** Nail starter board across the top of all rafters hanging over the edge of the house. Continue nailing starter boards up the rafters about 1 foot past the edge of the house.

**3.** Once the starter board has been nailed in place, work your way up the rafters with 1 by 4 strips. Space the strips the width of one board except for the final 3 feet of the rooftop, which should be covered solidly with 1 by 4s. This will make it easier to find nailing space when you adjust the last two or three rows of shakes or shingles.

**4.** If you are putting down an aluminum panel roof, you can space the 1 by 4s every 2 to 4 feet, depending on the pitch of your roof and expected snowloads. Check your local building codes.

# ROOFING

## Roofing Materials

When you select roofing material, consider the aesthetic appearance, durability, and cost. A roof of aluminum panels is short on appearance but long on durability; a shake roof is beautiful but costly; and asphalt shingles fall somewhere in between.

### Aluminum Panels

These lightweight panels come in lengths up to 16 feet, can be installed quickly, and will probably outlast the building you use them for. They are great for mountain cabins, barns, or workshops. Each panel is 26 inches wide to provide a 1-inch overlap on each side for rafters that are spaced 2 feet apart. Use aluminum roofing nails with a neoprene washer under the nail head. Panels are most easily cut with a carborundum blade on a circular saw, which is designed to cut light metals.

### Asphalt Shingles

These are widely used because of their relatively good appearance, reasonable cost, and long life (about 20 years). Before putting down asphalt shingles, cover the roof stripping with 30-pound roofing felt, which is heavy paper saturated with asphalt. It is laid horizontally across the roof with a 2- to 6-inch overlap. The last piece is folded over the ridge. In areas with heavy snow, put down a 36-inch-wide strip next to the eaves so that water will not be forced back under the shingles into the house when ice dams form. Use self-sealing shingles that have a strip of asphalt cement under each tab for wind resistance. These go down in a manner similar to shakes and shingles, with a double course on the first row.

### Shakes and Shingles

Shingles are machine cut and smooth on both sides, while shakes are split or half-sawn (sawn on one side). Both are generally made from red cedar and provide a beautiful, long-lasting roof. However, they are both highly susceptible to fire from chimney sparks. Some building codes specify that shakes and shingles must be treated with a fire retardant. If you can put down a shake roof, which is discussed in detail here, you will be able to put down the other styles quite easily.

Roofing felt is generally not required under shingles. However, it is recommended under shakes because they are more irregular, which means that wind and water can work through the cracks more easily.

Use 30-pound roofing felt that has guidelines marked on it to keep the shingle course straight. It should also have lines that indicate the precise amount of shake overlap. The felt is lightly tacked in place with a staple gun. A final strip is folded over the ridgeboard.

Use two 6d galvanized nails to nail down each shake (or two 4d galvanized nails for each shingle), regardless of size. A shingling hatchet is the right tool for this job, since you can use it to drive the nails, split the shakes to fit, and set your required shingle exposure (overlap). See page 61 for details on this tool.

How much exposure you should allow is determined by the pitch of your roof and the length of your shakes or shingles. You should discuss this question with the person you buy the shakes from as well as checking your local building codes. As a rule of thumb, you can consider 5½ inches a maximum exposure for shingles and 7½ inches

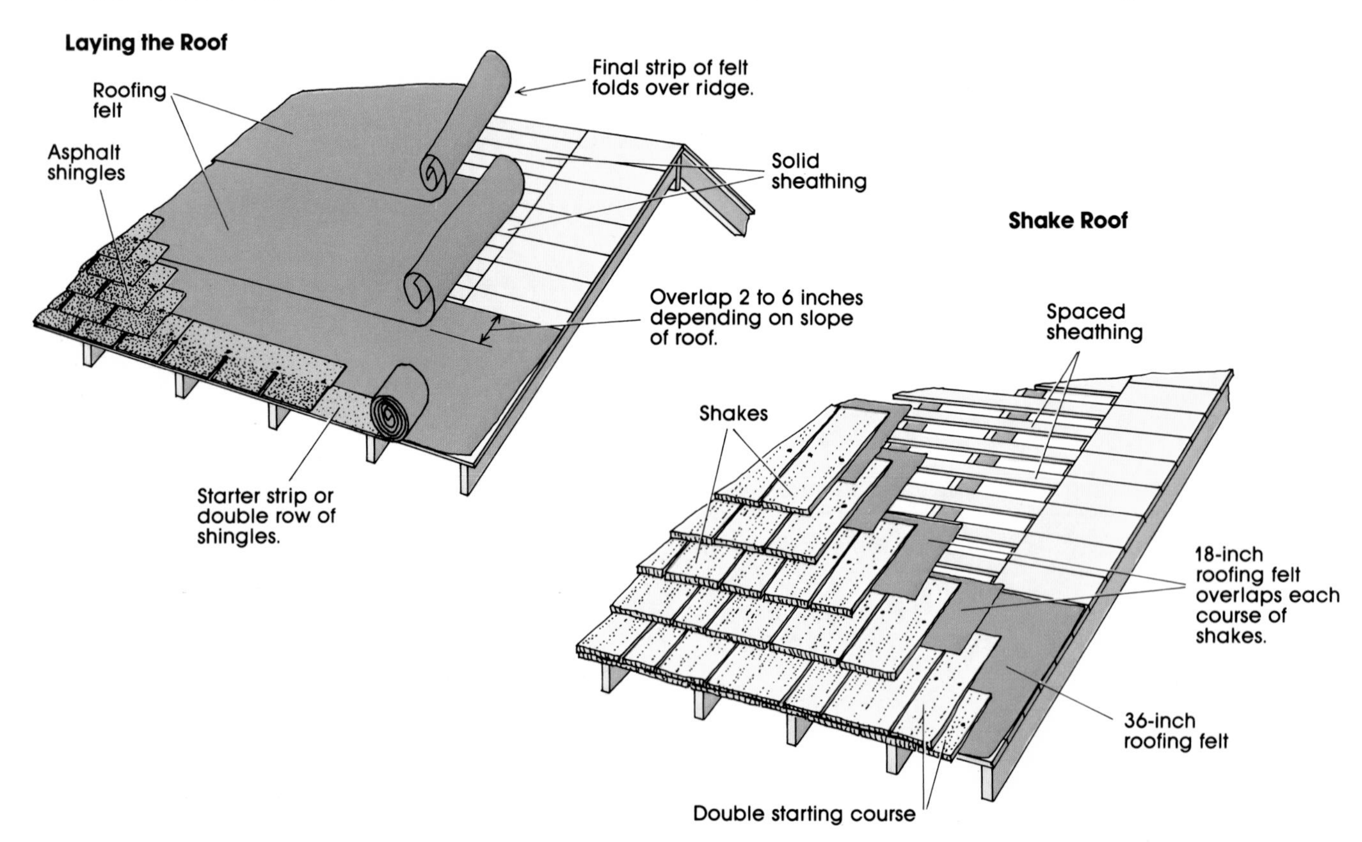

a maximum for shakes.

Shakes and shingles are sold by the bundle, and four bundles cover a *square,* which is 100 square feet. The surest way to estimate your needs is to give the exact dimensions of your roof to the shake supplier. Be sure to include the lengths of all ridges on the roof, since they require specially constructed ridge shingles.

## Laying the Shake Roof

The first row of shakes or shingles is always doubled to reduce the chance of ice dams pushing water back into the house. The butt ends should overhang the fascia board by the depth of your first knuckle. When you put down the first course, stop just short of driving the nails all the way through the starter boards, then bend the tops over. This will prevent the ends of the nails from showing under the eaves. If some show anyway, you can clip them off later.

With the first course, slip the upper end of the shake or shingle under the roofing felt. The second, or doubled, course lies on top of the felt. After that, all courses have the tops slipped under the felt. If the job is done right, no shingle ends should be seen on the underside of the roof, just the roofing felt.

As you work your way up, select your shakes by widths so that you overlap each crack by about 1½ inches. The butt of each shingle should overlap the nails by about ¾ inch.

When you nail down the shakes or shingles, it's important to leave about a 1⁄16- to 1⁄8-inch space between them. This gives them room to expand when they're wet and prevents them from buckling.

Shakes and shingles will form a pyramid shape as you work three or four rows at a time. Follow the guidelines on the roofing felt or snap your own with a chalkline.

The bundles should be scattered evenly on the roof so you won't have to go far each time you need a new bundle. Never open more than one bundle at a time or they may blow all over the place.

Work with particular care at the ends of the roof. The shingles should overhang the barge rafter about the depth of your first knuckle. Use that as a guide and select shakes with straight edges so the finished result is smooth and straight.

When you reach the final two rows, the shingle ends will protrude up in the air above the ridgeboard. When you're finished, snap a chalkline that falls over the center of the ridgeboard and trim the ends with a circular saw. This is faster, easier, and more accurate than trying to trim each shingle before you nail it down.

### Trimming in the Valleys

To quickly find the angle needed along the valley flashing, first snap a chalkline on each side of the flashing 2 inches back from the center.

As the last shingle on each course reaches this line, place it in position and mark at the top and bottom where it overhangs the line. Cut from point to point with your power saw. The result will be a smooth, straight line of shakes that leaves 2 inches of valley flashing exposed on each side.

**Trimming Shakes at the Ridge**

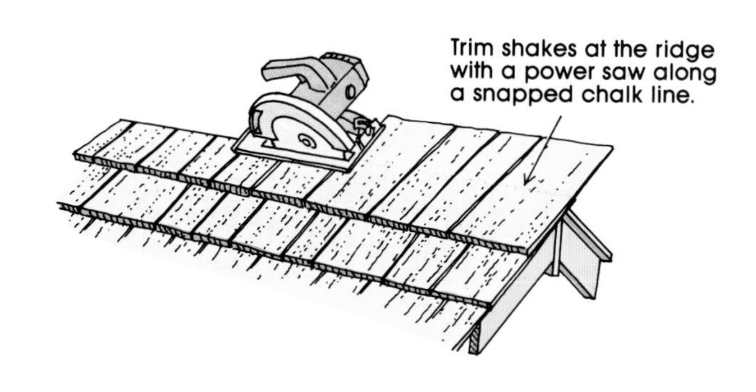


**Ridge Shakes**

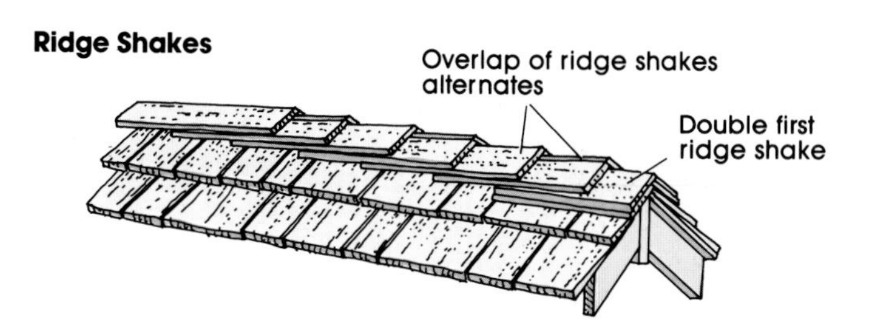


### Ridge Shakes and Shingles

These specially cut shakes or shingles are fabricated so that each successive overlap is opposite the previous one (see illustration) to prevent leaks.

The first two, at the end of the gable, are doubled. Overlap each section by 6 inches and nail them in place with two 6d nails to a side. Nail far enough back so that each overlap covers the nails.

You can put the ridge cap down in one continuous row or work from both ends and meet in the middle. In the latter case, double up the two middle pieces.

### Flashing

Flashing is designed to keep a building free of leaks. It is required around chimneys, vent pipes, and roof valleys. If you live in an area where ice dams build up on the roof eaves in the winter, you need the additional protection of roof felt from the eaves to 2 or 3 feet up the roof. (The installation of window flashing is described on page 103. For flashing repairs, see Ortho's book *Basic Home Repairs.*)

### Flashing Materials

Flashing is made from several different types of light metal, plastic, or asphalt-soaked paper such as roofing felt. No matter what type you select, don't try to get by with the cheapest material. If it rusts or rots out, it will cost you more to replace it than to have bought good material in the first place. And that cost doesn't include the mental anguish over the damaging leaks in the house.

**Metal flashings** include copper, lead, aluminum, stainless steel, galvanized steel, and terneplate. Of all of these, galvanized steel and terneplate (sheet steel that is coated with a lead-tin alloy) are the most widely used because of their quality and relatively low cost. Terneplate is prob-

# DOORS & WINDOWS

ably the best choice because it does not corrode like galvanized steel. Galvanized steel cannot be set in mortar or placed against masonry walls because the alkalies in the mortar will corrode it. The same is true with aluminum, unless it is first coated with asphalt. However, as long as galvanized steel is kept away from masonry, it is an excellent, long-lived flashing.

**Plastic flashings** are gaining in popularity because they are cheap, waterproof, and easy to handle.

**Asphalt-soaked flashings** are widely used around windows, as an undercoating on the roof before applying the roofing material, and to wrap a house for additional protection against moisture and wind penetration. They are inexpensive but tend to become brittle and crack as they age. Asphalt-soaked flashing should not be used where it will be exposed to the elements.

### Chimney Flashing

In order to get a tight seal between the chimney and the roof, chimney flashing must be placed in two stages before the roofing material is put down. The first, or base, layer of flashing is cut to fit around the chimney and to extend out on the roof at least 12 inches. The second, or cap, layer is embedded in the mortar of the chimney and caulked.

### Vent Pipe Flashing

Vent pipe flashing can often be purchased as a precut piece of metal that slips over the vent pipe. It has wide flanges at the base that extend out onto the roof to be covered by the roofing material.

### Pitched Roof Flashing

This type of flashing is used where a pitched roof line meets an existing wall, such as when a room is added on to the side of a house. Like chimney flashing, it is placed in two stages, with the cap flashing embedded in the mortar of the house siding (whether brick or stucco) or slipped under the wood siding.

### Valley Flashing

Valley flashing is placed where two downsloping roofs meet to form a valley. The flashing extends out under the roofing material for a distance specified by local building codes. The standard distance from the roofing material to the center of the valley flashing is 2 inches, but check your local codes.

The most commonly used valley flashing is called W flashing. The small ridge in the center gives it additional rigidity. It comes in various widths and should be selected to correspond with your local code requirements.

**Installing a Door**

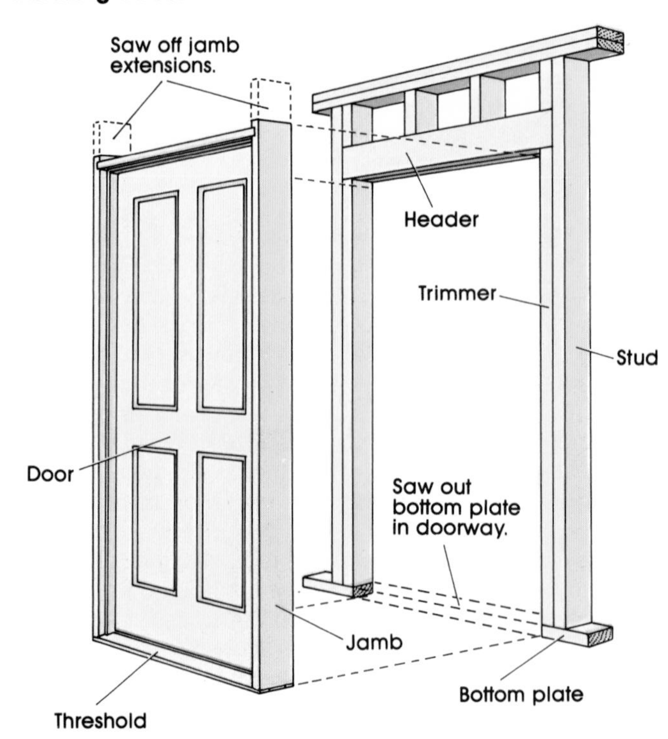


## Installing Doors

Doors and windows normally go in before the siding so the siding can be cut to fit accurately around them.

The easiest way to install a door is to buy one that is prehung—in other words, a door that is already hinged to the jamb. You just slip the jamb into the opening and nail it in place.

In actuality, it is not quite that simple. Installing a door takes care and patience. Here are the basic steps:

**1.** Cut out the sole plate from the door opening with a handsaw.

**2.** Trim off the protective ends on the doorjamb and remove any bracing.

**3.** Put the prehung unit into the door opening. Tap in four or five pieces of shingle along each side, between the jamb and the trimmer stud, to make sure the unit is plumb and square. Check repeatedly with your level.

**4.** Make sure the unit is centered in the opening so that the jamb protrudes the proper amount on both sides where the siding and interior walls will fit against it.

**5.** Use 10d casing nails to fasten the jamb to the trimmer

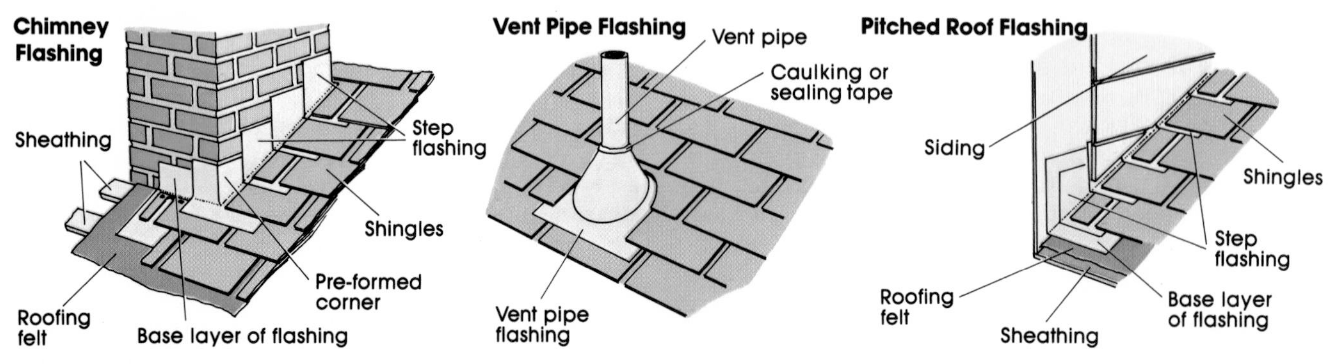

**Installing a Door**

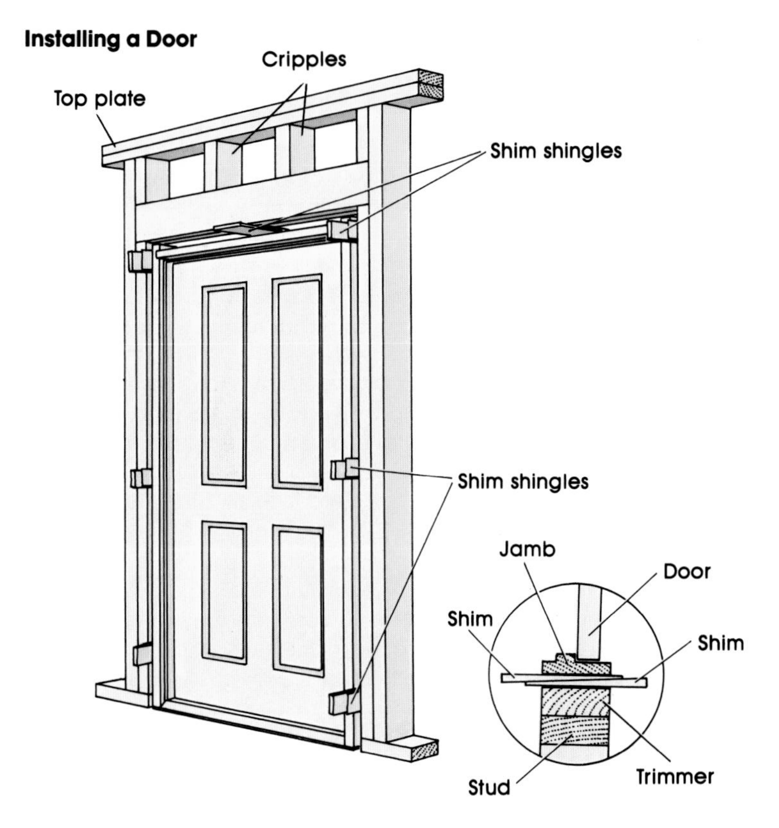


**Installing a Window**

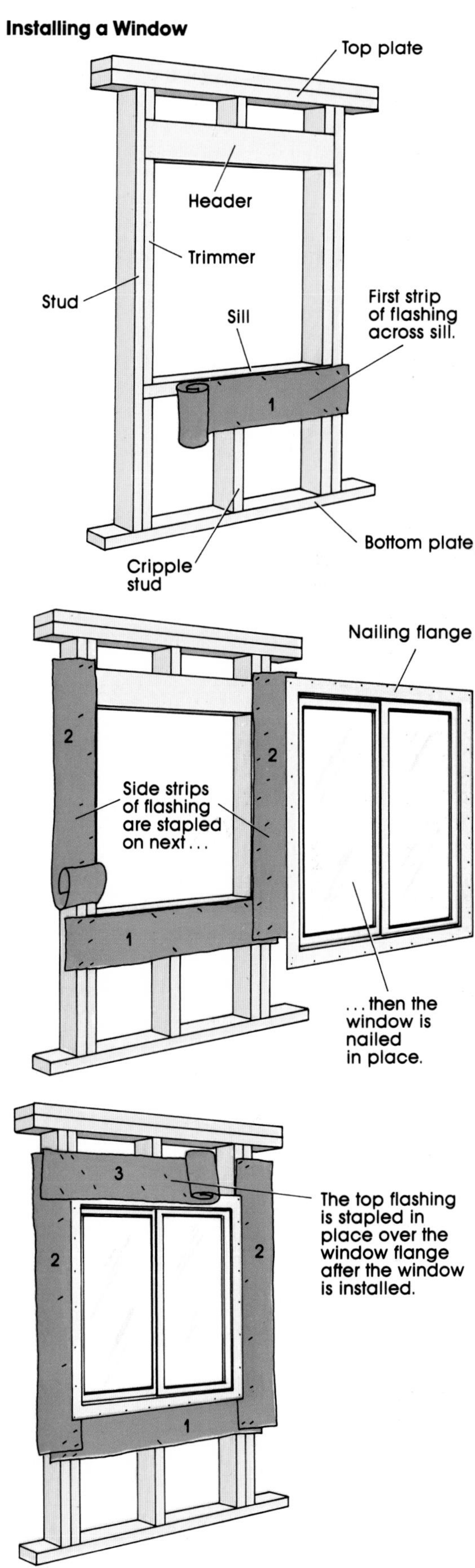


stud. Countersink the nails and fill the holes with wood putty.

**6.** Cut the protruding ends of the shim shingles flush with the trimmer studs.

**7.** It is not necessary to nail down the doorsill, which is part of the prehung door package. If you don't nail it down, the sill will be easier to replace should it become worn or broken.

## Installing Windows

Aluminum frame windows that take only a few minutes to install are now widely used in construction. But first the outside of the window opening must be covered with flashing paper. This is put on in a way that prevents water from leaking around the siding and past the window.

First, run a strip across the windowsill, the top edge of the paper flush with the top of the sill. Then cover both sides, with the bottom edges overlapping the first strip. The last strip goes on after the window is installed and covers the top flange of the window frame.

After the window flashing is stapled up, slip the window into the opening. The window is installed with the screen and the drain holes along the bottom edge on the outside. One person works inside with a pry bar and a level to make sure the window is square and level. While the person inside holds the window in place, the person outside drives 8d nails through the flanges into the trimmer studs and header. Drive just three or four nails part way to hold it, then check that the window still opens and closes smoothly. If it does, finish nailing, spacing the nails about 12 inches apart.

The last step is to staple up the top piece of window flashing.

# SIDING

Because of the wide variety of siding material available, your choice will depend on how formal you want it to look, how much time you have left, and your pocketbook.

Plywood siding is one of the least expensive types, goes up rapidly, adds considerable rigidity to the walls, and is available in several different designs. It comes in sheets 4 feet wide and lengths of 8, 9, 10, or 12 feet. Generally, ⅜-inch siding is sufficiently thick and is light enough for one person to handle.

If the building is constructed with standard 92¼-inch studs, you must use 9-foot lengths of plywood siding rather than 8-foot lengths. This is because the height of the wall with the sole plate and double top plates is 96¾ inches high. The extra ¾ inch gives you room to put up sheetrock or paneling inside, but you must use longer plywood siding outside to cover the wall and part of the foundation.

Before putting up the siding, you should consider wrapping the building with building paper to make it more airtight. Remember to check your local building codes, since building paper may be required.

Start putting the plywood panels up at one corner. They are normally placed vertically, but in earthquake-prone areas, you may be required to place them horizontally. This way they overlap more studs at a time and make the walls even stronger. Use 6d galvanized nails for siding that is ½ inch thick or less.

The leading edge of the first panel should fall exactly down the center of a stud. If this results in the panel being slightly off at the wall corner, let it be. You can hide it with the trim material. But where it falls on that stud at the 4-foot mark, it must be perfectly aligned and straight. If the stud is badly out of line, nail some backing on it and nail the panel to the backing.

This is where carefully placed studs pay off. Sloppy work means that the siding panels won't fit properly, and you will have to do a lot of extra work to make them fit.

When you make the cutouts around doors and windows, allow yourself ½ inch extra space. This gives you ¼ inch room at the top, bottom, or side for adjustments. The small gap will be covered with trim.

When the walls are covered, go to work on the gable ends. To determine angles, measure the width along the bottom and then the height. The final angle of the roof line, the hypotenuse, is found by connecting the ends of the height and length measurements.

Where a section of plywood rests on the lower wall panels, such as between the gable ends and the main walls, insert strips of metal flashing called *Z bars* to prevent rain from working through the joint.

After the siding is completed, caulk all joints between the panels and around doors and windows to help weatherproof the house.

The protective and decorative stain of your choice is applied next. It should be put on with a brush in order to penetrate the wood properly. The second choice is to use a deep-nap roller. Don't use a spray gun—the stain won't penetrate properly.

## Trimming

Trim material for plywood paneling is often 1-by-2 or 1-by-4 pine. A 1-by-2 strip called a batten is nailed over each joint between panels. Strips are also overlapped at each corner of the house. Around doors and windows, decorative angles may be cut at the ends of the trim. The trim piece above the window should extend out over the side trim to prevent water from running in.

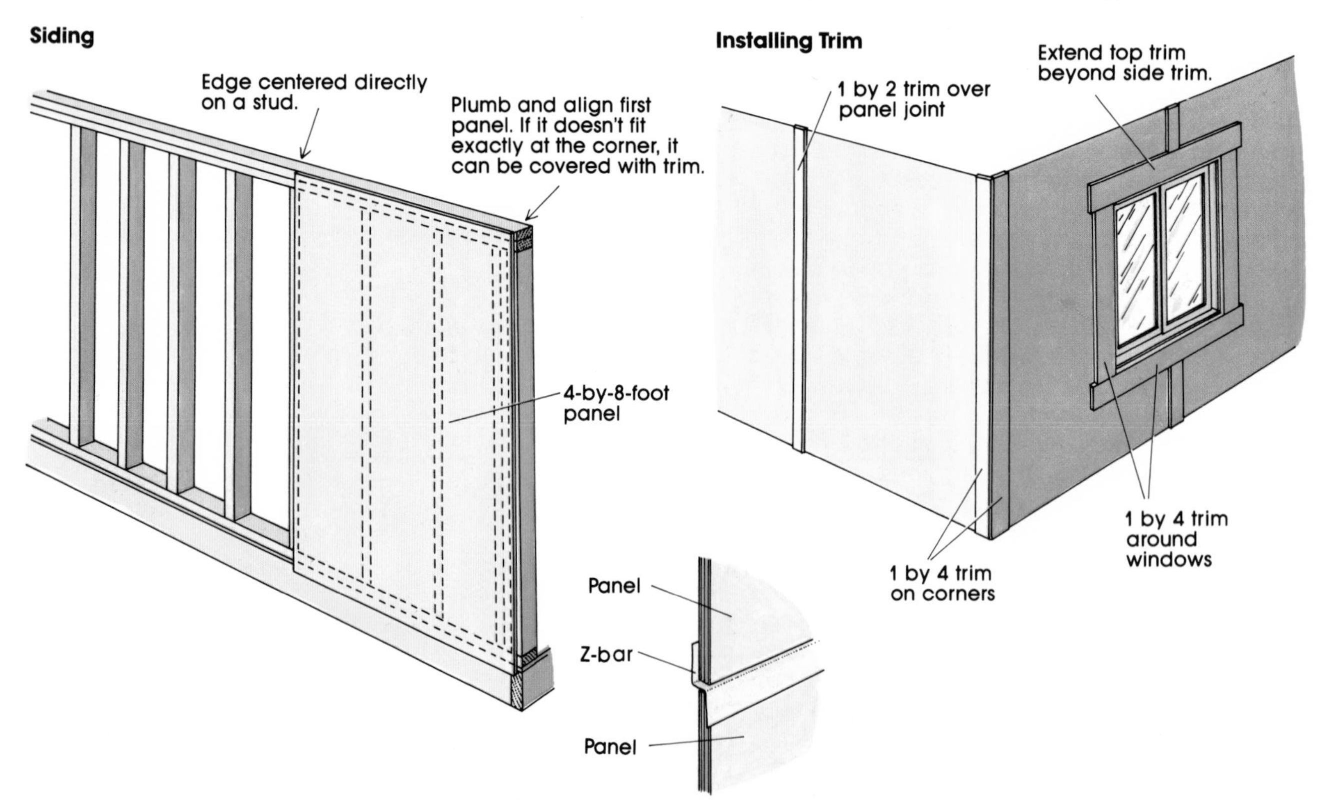

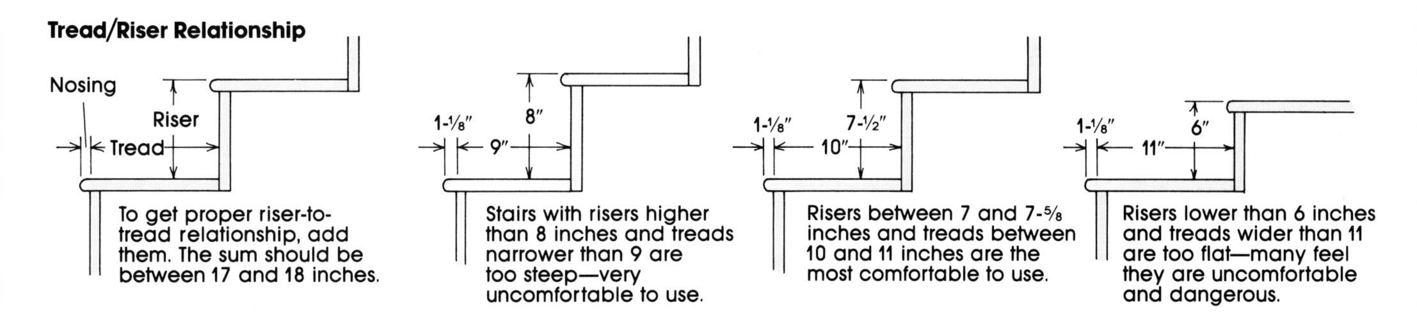


Stairs range from the intricately beautiful to the simply functional. They also require some mathematics to make them come out right, so sharpen your pencil or get your calculator ready.

There is a given relationship between the width of the tread and the height of the riser. When these two are added together, the total should be between 17 and 18 inches. Thus if the riser is 7 inches, the tread will be 11 inches. Tread widths should be between 10 and 11 inches and risers between 7 and 8. The calculation of these figures depends on the rise of the stairs and angle of the stringers.

Stringers should always be made from 2 by 12s for strength and to simplify laying them out with the framing square.

Stairways should generally be wide enough for two people to pass each other on them, with 3 feet being a common width for utility stairs. Make them 4 feet wide or more on decks and patios for an open and pleasing appearance. You should use a central supporting stringer for stairs more than 3 feet wide.

The simplest type of stairway consists of two stringers with cleats nailed on and treads nailed to the cleats. The appearance can be improved somewhat by dadoing the stringers and then gluing and nailing the tread in place. But the safest and strongest stairs are made with a cut-out sawtooth pattern. This type will be discussed here.

The angle of the stairway is determined in part by how much room you have. However, it should never be less than 20 degrees or more than 50 degrees. An angle between 30 and 35 degrees is ideal.

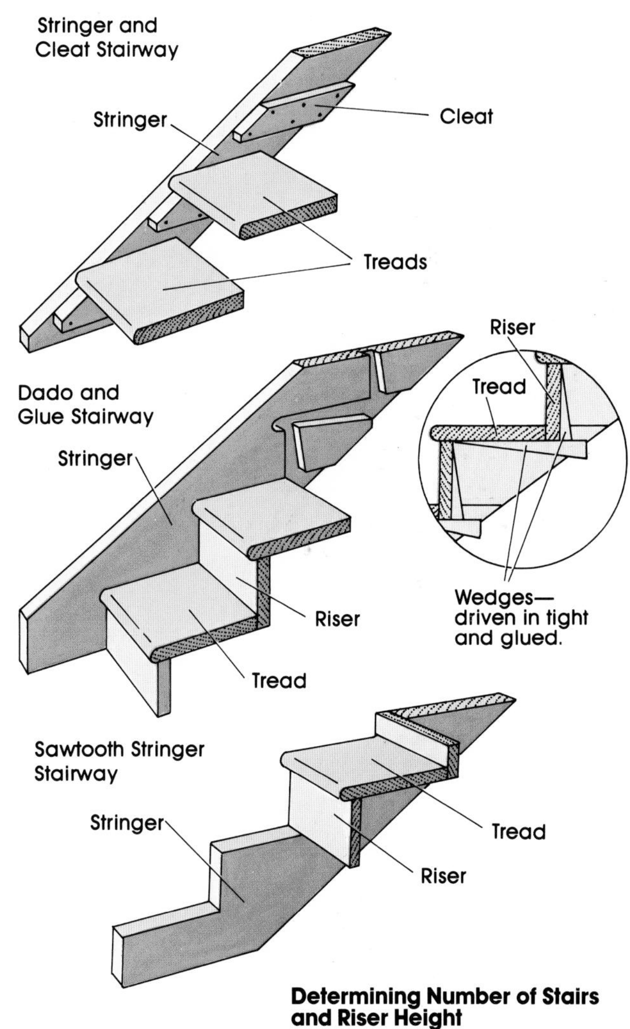


## Determining the Number of Steps

One way to determine the number of steps required using a minimum amount of mathematics is with a *story pole*, which is simply a straight piece of 1 by 2 or 1 by 4. First, place the pole vertically and mark the location of the floor above. If the flooring isn't laid yet, add that additional height to the story pole. Always calculate from finish floor to finish floor.

Now, with a pair of wing dividers set 7 inches apart, step off the distance from the bottom (B) to the top (A). The object is to divide this distance into equal parts. It's not likely you will come out evenly on the first try. If the remainder is less than 3½ inches, set the dividers wider than 7 inches; if the remainder is more than 3½ inches, set the dividers to less than 7 inches. When, through trial and error, you have divided the pole into equal parts, you have found how high the risers should be and how many there are. To find the width of the tread, subtract the riser height

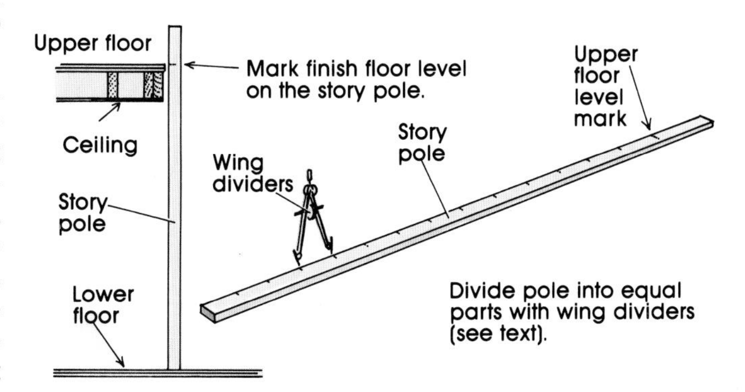

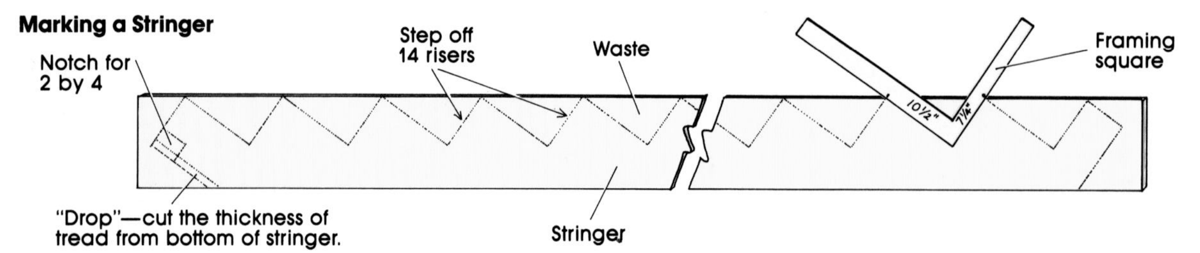


from 17½ (because the sum of one riser and one tread should fall between 17 and 18).

To find the number of risers mathematically, divide the total rise in inches by 7 (drop any fractions). Then, to find the exact height of each riser, divide the total rise by the number of risers.

For example if the total rise is 9 feet (108 inches), first divide that by 7 to get 15.42. Dropping the fraction, we have 15 risers. Dividing the total rise of 108 inches by 15 gives us 7.2 or 7¼ inches as the height of each riser. The tread width is 17½ minus 7¼, or 10¼ inches.

To find the total run (length) of the stringers, multiply the width of the tread by one less than the number of risers (because there is always one less tread than the number of risers). Continuing this example, the length of the stringer is

$$10\frac{1}{4} \times 14 = 143\frac{1}{2} \text{ inches}$$

Got that?

## Layout of Stringers

With your figures close at hand, use your framing square to mark off the cuts on the 2 by 12. For the example given above, you would set the framing square on the 2 by 12 and measure off a 10¼-inch tread with the blade and a 7¼-inch riser with the tongue.

Start at the bottom of the stringer and mark out the first step. Now draw line AD at right angles to AB. The length of AD is the same as your riser unit. Line DE, at right angles to AD, is the foot or base of the stringer.

Continue stepping off the stairs on the stringer to the top, and make the cuts. Finally, from the bottom trim off the thickness of one tread. This adjusts it to the floor level, which has no tread. Put the stringer up to see if it fits. If it does, use it as a pattern for the other stringer.

Secure the stringers in place at the top by hanging them from joist hangers (see page 85). At the bottom, nail down a 2 by 4 and notch the foot of the stringers to fit over it, as illustrated, then toenail the stringers to the 2 by 4.

When you nail on the treads and risers, put the riser on first and then the tread. The tread should extend out over the riser about 1 inch.

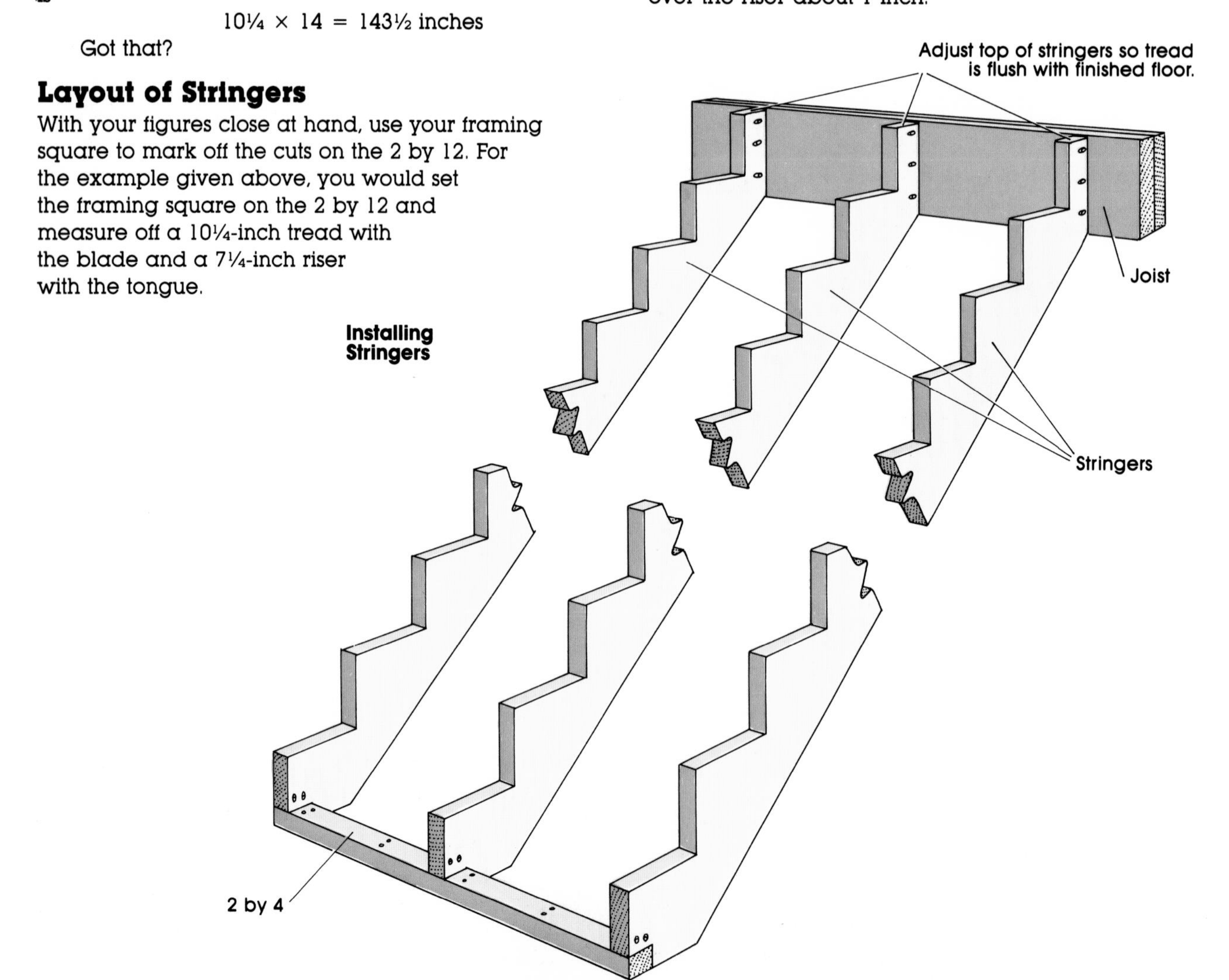

## Constructing Railings

Stairs should have railings on both sides. The standard railing height is 32 or 33 inches from the top of a riser to the top of the handrail. Always measure from the same point at the top and bottom of the stairs so the rail will be parallel to the stairs. The railings should have smooth, rounded corners and should be free of splinters.

Posts should be made from 4 by 4s or something close to that size. If you use 4-by-4 posts, notch one side of each post at the bottom ¾ inch deep and fit this against the stringer before putting down the treads. (The tread should be cut to fit around the post.) This notch, with its lip resting on the stringer's tread notch, will increase the post's stability. Use two bolts to fasten the post to the stringer. Space the posts 3 to 6 feet apart so they are equally spaced down the stairs. Use 2 by 6s for the top railing. These provide an aesthetically pleasing overhang on each side of the post. For the middle railing, use 2 by 4s toenailed between the posts. If you are concerned about children falling between the railings, nail 2 by 2s about 8 inches apart between the top and middle railings.

**Constructing a Railing**

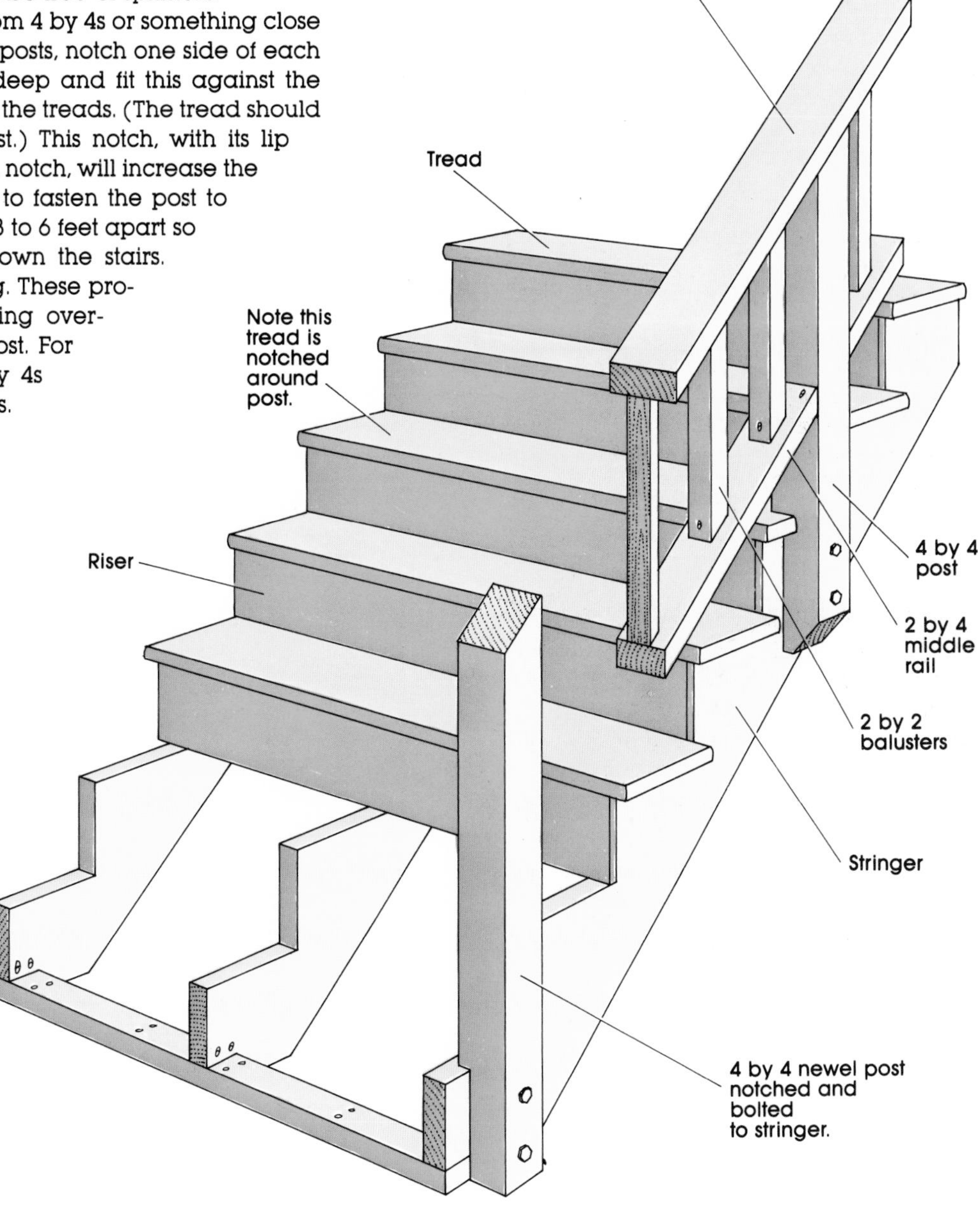


## What Next?

The building is up, the roof is on, the windows and doors are installed, and the siding and trim are finished. You have broken out the grog for your fellow workers, and you feel like the job is almost finished. If you have built a simple shop, garage, or barn, then the job is indeed just about over except for some electrical wiring.

But if you have built a house, your job is less than halfway done. You'll need to run all the wiring and hook it into the outlets and switches; install wall and ceiling insulation; put up the sheetrock on the walls and ceiling, and tape and plaster it; attach the baseboards and all the trim molding; paint everything; put down finish floors and carpeting; install the kitchen and bathroom fixtures; build and install all the cabinets—all right, enough—it's not necessary to spoil your present sense of success.

There's a lot of work ahead, but go at it with confidence. Take each job one step at a time, just as you did with the construction, and you'll be finished before you know it.

# SPAN CHARTS

Joists and rafters must bear the weight of people and furniture (live loads) and of the structure itself (dead loads), and are, therefore, subject to stress and bending. The wood fibers in different grades of lumber have varying levels of stress capabilities. Span charts were developed to help builders quickly determine how long a joist or rafter can be and still withstand common loads.

While there are many more charts available, these charts show common construction lumber species, grades and sizes, and common joist and rafter spacing. Before you begin any major construction project, you should check your local building codes which will include charts for all the possible variations.

**How to read Chart 1**

**1.** In the left-hand columns, find the species and grade of lumber you are using.

**2.** Move across the line to the next columns: *Floor or Ceiling Members* and *Roof Members*. If you live in snow country, you will refer to the numbers under "snow loading." Numbers in these columns represent the pounds per square inch that particular species of wood can bear. When you have found the appropriate stress number, go to Charts 2, 3, 4, and 5.

**How to read Charts 2, 3, 4, and 5**

**1.** In the left columns, find your lumber size and joist spacing (16" and 24" o.c. is most common).

**2.** Move across the columns on that line and locate the stress number you found in Chart 1. If your number falls in between the numbers listed in Chart 2, find the closest number that is *lower* than yours. The numbers shown above the stress number (7-9, 8-0, 8-4, etc.) are allowable spans in feet and inches (7′ 9″, 8′ 0″, etc.) for that wood species and size.

**CHART 1*– WORKING STRESSES FOR JOISTS AND RAFTERS – VISUAL GRADING – These $F_b$ values are for use where repetitive members are spaced not more than 24 inches. For wider spacing, the $F_b$ values should be reduced 13 percent. Values are for surfaced dry or surfaced green lumber, except for southern pine as indicated. Values apply at 19 percent maximum moisture content in use.**

| SPECIES | GRADE | SIZE | ALLOWABLE UNIT STRESS IN BENDING $F_b$ | | | MODULUS OF ELASTICITY $E$ $1 \times 10^6$ psi |
|---|---|---|---|---|---|---|
| | | | FLOOR OR CEILING MEMBERS | ROOF MEMBERS | | |
| | | | | SNOW LOADING | NO SNOW LOADING | |
| BALSAM FIR<br>EASTERN WHITE PINE<br>EASTERN WHITE PINE (NORTH)<br>ENGELMANN SPRUCE—ALPINE FIR—LODGEPOLE PINE<br>RED PINE | Construction | 2×4 | 800 | 920 | 1000 | 1.0 (1.2) |
| | Standard | 2×4 | 450 | 520 | 560 | 1.0 (.9) |
| | Studs | 2×4 | 600 | 690 | 750 | 1.0 (.9) |
| | No. 1 & Appearance | 2×6 & Wider | 1150 | 1320 | 1440 | 1.2 (1.2) |
| | No. 2 | 2×6 & Wider | 950 | 1090 | 1190 | 1.1 (1.1) |
| | No. 3 | 2×6 & Wider | 550 | 630 | 690 | 1.0 (.9) |
| CALIFORNIA REDWOOD | Construction | 2×4 | 950 | 1090 | 1190 | .9 |
| | Standard | 2×4 | 550 | 630 | 690 | .9 |
| | Studs | 2×4 | 700 | 800 | 880 | .9 |
| | No. 1 | 2×6 & Wider | 1700 | 1960 | 2120 | 1.4 |
| | No. 1, Open grain | 2×6 & Wider | 1350 | 1550 | 1690 | 1.1 |
| | No. 2 | 2×6 & Wider | 1400 | 1610 | 1750 | 1.3 |
| | No. 2, Open grain | 2×6 & Wider | 1100 | 1260 | 1370 | 1.0 |
| | No. 3 | 2×6 & Wider | 800 | 920 | 1000 | 1.1 |
| | No. 3, Open grain | 2×6 & Wider | 650 | 750 | 810 | .9 |
| DOUGLAS FIR—LARCH<br>DOUGLAS FIR—LARCH (NORTH) | Construction | 2×4 | 1200 | 1380 | 1500 | 1.5 |
| | Standard | 2×4 | 675 | 780 | 840 | 1.5 |
| | Studs | 2×4 | 925 | 1060 | 1160 | 1.5 |
| | No. 1 & Appearance | 2×6 & Wider | 1750 | 2010 | 2190 | 1.8 |
| | Dense No. 2 | 2×6 & Wider | 1700 | 1960 | 2120 | 1.7 |
| | No. 2 | 2×6 & Wider | 1450 | 1670 | 1810 | 1.7 |
| | No. 3 | 2×6 & Wider | 850 | 980 | 1060 | 1.5 |
| DOUGLAS FIR SOUTH | Construction | 2×4 | 1150 | 1320 | 1440 | 1.1 |
| | Standard | 2×4 | 650 | 750 | 810 | 1.1 |
| | Studs | 2×4 | 875 | 1010 | 1090 | 1.1 |
| | No. 1 & Appearance | 2×6 & Wider | 1650 | 1900 | 2060 | 1.4 |
| | No. 2 | 2×6 & Wider | 1350 | 1550 | 1690 | 1.3 |
| | No. 3 | 2×6 & Wider | 800 | 920 | 1000 | 1.1 |
| EASTERN HEMLOCK—TAMARACK<br>EASTERN HEMLOCK (NORTH) | Construction | 2×4 | 1050 | 1210 | 1310 | 1.0 |
| | Standard | 2×4 | 575 | 660 | 720 | 1.0 |
| | Studs | 2×4 | 800 | 920 | 1000 | 1.0 |
| | No. 1 & Appearance | 2×6 & Wider | 1500 | 1720 | 1880 | 1.3 |
| | No. 2 | 2×6 & Wider | 1200 | 1380 | 1500 | 1.1 |
| | No. 3 | 2×6 & Wider | 725 | 830 | 910 | 1.0 |
| COAST SPECIES[3]<br>EASTERN SPRUCE | Construction | 2×4 | 875 | 1010 | 1090 | 1.1 (1.2) |
| | Standard | 2×4 | 500 | 580 | 620 | 1.1 (1.2) |
| | Studs | 2×4 | 675 | 780 | 840 | 1.1 (1.2) |
| | No. 1 & Appearance | 2×6 & Wider | 1250 | 1440 | 1560 | 1.4 (1.5) |
| | No. 2 | 2×6 & Wider | 1000 | 1150 | 1250 | 1.2 (1.4) |
| | No. 3 | 2×6 & Wider | 600 | 690 | 750 | 1.1 (1.2) |
| HEM—FIR<br>HEM—FIR (NORTH)<br>MOUNTAIN HEMLOCK—HEM—FIR[4] | Construction | 2×4 | 975 | 1120 | 1220 | 1.2 (1.0) |
| | Standard | 2×4 | 550 | 630 | 690 | 1.2 (1.0) |
| | Studs | 2×4 | 725 | 830 | 910 | 1.2 (1.0) |
| | No. 1 & Appearance | 2×6 & Wider | 1400 | 1610 | 1750 | 1.5 (1.3) |
| | No. 2 | 2×6 & Wider | 1150 | 1320 | 1440 | 1.4 (1.1) |
| | No. 3 | 2×6 & Wider | 675 | 780 | 840 | 1.2 (1.0) |

## CHART 1, CONTINUED — WORKING STRESSES FOR JOISTS AND RAFTERS

| SPECIES | GRADE | SIZE | ALLOWABLE UNIT STRESS IN BENDING $F_b$ | | | MODULUS OF ELASTICITY $E$ $1 \times 10^6$ psi |
|---|---|---|---|---|---|---|
| | | | FLOOR OR CEILING MEMBERS | ROOF MEMBERS | | |
| | | | | SNOW LOADING | NO SNOW LOADING | |
| **IDAHO WHITE PINE WESTERN WHITE PINE WHITE WOODS (MIXED SPECIES)**[5] | Construction | 2×4 | 775 | 890 | 970 | 1.2 (.9) |
| | Standard | 2×4 | 425 | 490 | 530 | 1.2 (.9) |
| | Studs | 2×4 | 600 | 690 | 750 | 1.2 (.9) |
| | No. 1 & Appearance | 2×6 & Wider | 1100 | 1260 | 1370 | 1.4 (1.1) |
| | No. 2 | 2×6 & Wider | 925 | 1060 | 1160 | 1.3 (1.0) |
| | No. 3 | 2×6 & Wider | 550 | 630 | 690 | 1.2 (.9) |
| **NORTHERN PINE** | Construction | 2×4 | 950 | 1090 | 1190 | 1.1 |
| | Standard | 2×4 | 525 | 600 | 660 | 1.1 |
| | Studs | 2×4 | 725 | 830 | 910 | 1.1 |
| | No. 1 & Appearance | 2×6 & Wider | 1400 | 1610 | 1750 | 1.4 |
| | No. 2 | 2×6 & Wider | 1100 | 1260 | 1380 | 1.3 |
| | No. 3 | 2×6 & Wider | 650 | 750 | 810 | 1.1 |
| **PONDEROSA PINE— SUGAR PINE— LODGEPOLE PINE** | Construction | 2×4 | 825 | 950 | 1030 | 1.0 |
| | Standard | 2×4 | 450 | 520 | 560 | 1.0 |
| | Studs | 2×4 | 625 | 720 | 780 | 1.0 |
| | No. 1 & Appearance | 2×6 & Wider | 1200 | 1380 | 1500 | 1.2 |
| | No. 2 | 2×6 & Wider | 975 | 1120 | 1220 | 1.1 |
| | No. 3 | 2×6 & Wider | 575 | 660 | 720 | 1.0 |
| | No. 1 & Appearance | 2×6 & Wider | 1300 | 1500 | 1620 | 1.5 |
| | No. 2 | 2×6 & Wider | 1050 | 1210 | 1310 | 1.3 |
| | No. 3 | 2×6 & Wider | 600 | 690 | 750 | 1.2 |
| **SOUTHERN PINE (Surfaced dry)** | Construction | 2×4 | 1150 | 1320 | 1435 | 1.4 |
| | Standard | 2×4 | 675 | 775 | 840 | 1.4 |
| | Studs | 2×4 | 900 | 1035 | 1125 | 1.4 |
| | No. 1 | 2×6 & Wider | 1700 | 1955 | 2125 | 1.7 |
| | No. 1 Dense | 2×6 & Wider | 2000 | 2300 | 2500 | 1.8 |
| | No. 2 | 2×6 & Wider | 1400 | 1610 | 1750 | 1.6 |
| | No. 2 Dense | 2×6 & Wider | 1650 | 1900 | 2060 | 1.7 |
| | No. 3 | 2×6 & Wider | 800 | 920 | 1000 | 1.4 |
| | No. 3 Dense | 2×6 & Wider | 925 | 1060 | 1155 | 1.5 |

## CHART 2 – FLOOR JOISTS — Allowable spans for 40 lbs. per sq. ft. live load.

| JOIST SIZE | JOIST SPACING | MODULUS OF ELASTICITY, $E$, IN 1,000,000 psi | | | | | | | | | | | | | |
|---|---|---|---|---|---|---|---|---|---|---|---|---|---|---|---|
| | | 0.8 | 0.9 | 1.0 | 1.1 | 1.2 | 1.3 | 1.4 | 1.5 | 1.6 | 1.7 | 1.8 | 1.9 | 2.0 | 2.2 |
| 2×6 | 16.0″ | 7-9<br>790 | 8-0<br>860 | 8-4<br>920 | 8-7<br>980 | 8-10<br>1040 | 9-1<br>1090 | 9-4<br>1150 | 9-6<br>1200 | 9-9<br>1250 | 9-11<br>1310 | 10-2<br>1360 | 10-4<br>1410 | 10-6<br>1460 | 10-10<br>1550 |
| | 24.0″ | 6-9<br>900 | 7-0<br>980 | 7-3<br>1050 | 7-6<br>1120 | 7-9<br>1190 | 7-11<br>1250 | 8-2<br>1310 | 8-4<br>1380 | 8-6<br>1440 | 8-8<br>1500 | 8-10<br>1550 | 9-0<br>1610 | 9-2<br>1670 | 9-6<br>1780 |
| 2×8 | 16.0″ | 10-2<br>790 | 10-7<br>850 | 11-0<br>920 | 11-4<br>980 | 11-8<br>1040 | 12-0<br>1090 | 12-3<br>1150 | 12-7<br>1200 | 12-10<br>1250 | 13-1<br>1310 | 13-4<br>1360 | 13-7<br>1410 | 13-10<br>1460 | 14-3<br>1550 |
| | 24.0″ | 8-11<br>900 | 9-3<br>980 | 9-7<br>1050 | 9-11<br>1120 | 10-2<br>1190 | 10-6<br>1250 | 10-9<br>1310 | 11-0<br>1380 | 11-3<br>1440 | 11-5<br>1500 | 11-8<br>1550 | 11-11<br>1610 | 12-1<br>1670 | 12-6<br>1780 |
| 2×10 | 16.0″ | 13-0<br>790 | 13-6<br>850 | 14-0<br>920 | 14-6<br>980 | 14-11<br>1040 | 15-3<br>1090 | 15-8<br>1150 | 16-0<br>1200 | 16-5<br>1250 | 16-9<br>1310 | 17-0<br>1360 | 17-4<br>1410 | 17-8<br>1460 | 18-3<br>1550 |
| | 24.0″ | 11-4<br>900 | 11-10<br>980 | 12-3<br>1050 | 12-8<br>1120 | 13-0<br>1190 | 13-4<br>1250 | 13-8<br>1310 | 14-0<br>1380 | 14-4<br>1440 | 14-7<br>1500 | 14-11<br>1550 | 15-2<br>1610 | 15-5<br>1670 | 15-11<br>1780 |
| 2×12 | 16.0″ | 15-10<br>790 | 16-5<br>860 | 17-0<br>920 | 17-7<br>980 | 18-1<br>1040 | 18-7<br>1090 | 19-1<br>1150 | 19-6<br>1200 | 19-11<br>1250 | 20-4<br>1310 | 20-9<br>1360 | 21-1<br>1410 | 21-6<br>1460 | 22-2<br>1550 |
| | 24.0″ | 13-10<br>900 | 14-4<br>980 | 14-11<br>1050 | 15-4<br>1120 | 15-10<br>1190 | 16-3<br>1250 | 16-8<br>1310 | 17-0<br>1380 | 17-5<br>1440 | 17-9<br>1500 | 18-1<br>1550 | 18-5<br>1610 | 18-9<br>1670 | 19-4<br>1780 |

## CHART 3 – CEILING JOISTS — Allowable spans for 10 lbs. per sq. ft. live load. (Drywall Ceiling)

| JOIST SIZE | JOIST SPACING | 0.8 | 0.9 | 1.0 | 1.1 | 1.2 | 1.3 | 1.4 | 1.5 | 1.6 | 1.7 | 1.8 | 1.9 | 2.0 | 2.2 |
|---|---|---|---|---|---|---|---|---|---|---|---|---|---|---|---|
| 2×4 | 16.0″ | 8-11<br>780 | 9-4<br>850 | 9-8<br>910 | 9-11<br>970 | 10-3<br>1030 | 10-6<br>1080 | 10-9<br>1140 | 11-0<br>1190 | 11-3<br>1240 | 11-6<br>1290 | 11-9<br>1340 | 11-11<br>1390 | 12-2<br>1440 | 12-6<br>1540 |
| | 24.0″ | 7-10<br>900 | 8-1<br>970 | 8-5<br>1040 | 8-8<br>1110 | 8-11<br>1170 | 9-2<br>1240 | 9-5<br>1300 | 9-8<br>1360 | 9-10<br>1420 | 10-0<br>1480 | 10-3<br>1540 | 10-5<br>1600 | 10-7<br>1650 | 10-11<br>1760 |
| 2×6 | 16.0″ | 14-1<br>780 | 14-7<br>850 | 15-2<br>910 | 15-7<br>970 | 16-1<br>1030 | 16-6<br>1080 | 16-11<br>1140 | 17-4<br>1190 | 17-8<br>1240 | 18-1<br>1290 | 18-5<br>1340 | 18-9<br>1390 | 19-1<br>1440 | 19-8<br>1540 |
| | 24.0″ | 12-3<br>900 | 12-9<br>970 | 13-3<br>1040 | 13-8<br>1110 | 14-1<br>1170 | 14-5<br>1240 | 14-9<br>1300 | 15-2<br>1360 | 15-6<br>1420 | 15-9<br>1480 | 16-1<br>1540 | 16-4<br>1600 | 16-8<br>1650 | 17-2<br>1760 |
| 2×8 | 16.0″ | 18-6<br>780 | 19-3<br>850 | 19-11<br>910 | 20-7<br>970 | 21-2<br>1030 | 21-9<br>1080 | 22-4<br>1140 | 22-10<br>1190 | 23-4<br>1240 | 23-10<br>1290 | 24-3<br>1340 | 24-8<br>1390 | 25-2<br>1440 | 25-11<br>1540 |
| | 24.0″ | 16-2<br>900 | 16-10<br>970 | 17-5<br>1040 | 18-0<br>1110 | 18-6<br>1170 | 19-0<br>1240 | 19-6<br>1300 | 19-11<br>1360 | 20-5<br>1420 | 20-10<br>1480 | 21-2<br>1540 | 21-7<br>1600 | 21-11<br>1650 | 22-8<br>1760 |
| 2×10 | 16.0″ | 23-8<br>780 | 24-7<br>850 | 25-5<br>910 | 26-3<br>970 | 27-1<br>1030 | 27-9<br>1080 | 28-6<br>1140 | 29-2<br>1190 | 29-9<br>1240 | 30-5<br>1290 | 31-0<br>1340 | 31-6<br>1390 | 32-1<br>1440 | 33-1<br>1540 |
| | 24.0″ | 20-8<br>900 | 21-6<br>970 | 22-3<br>1040 | 22-11<br>1110 | 23-8<br>1170 | 24-3<br>1240 | 24-10<br>1300 | 25-5<br>1360 | 26-0<br>1420 | 26-6<br>1480 | 27-1<br>1540 | 27-6<br>1600 | 28-0<br>1650 | 28-11<br>1760 |

**CHART 4 – LOW OR HIGH SLOPE RAFTERS** — Allowable spans for 30 lbs. per sq. ft. live load. (Supporting Drywall Ceiling)

| RAFTER SIZE | RAFTER SPACING | ALLOWABLE EXTREME FIBER STRESS IN BENDING $F_b$ (psi). | | | | | | | | | | | | | | |
|---|---|---|---|---|---|---|---|---|---|---|---|---|---|---|---|---|
| | | 500 | 600 | 700 | 800 | 900 | 1000 | 1100 | 1200 | 1300 | 1400 | 1500 | 1600 | 1700 | 1800 | 1900 |
| 2×6 | 16.0″ | 6-6<br>0.24 | 7-1<br>0.31 | 7-8<br>0.39 | 8-2<br>0.48 | 8-8<br>0.57 | 9-2<br>0.67 | 9-7<br>0.77 | 10-0<br>0.88 | 10-5<br>0.99 | 10-10<br>1.10 | 11-3<br>1.22 | 11-7<br>1.35 | 11-11<br>1.48 | 12.4<br>1.61 | 12.8<br>1.75 |
| | 24.0″ | 5-4<br>0.19 | 5-10<br>0.25 | 6-3<br>0.32 | 6-8<br>0.39 | 7-1<br>0.46 | 7-6<br>0.54 | 7-10<br>0.63 | 8-2<br>0.72 | 8-6<br>0.81 | 8-10<br>0.90 | 9-2<br>1.00 | 9-6<br>1.10 | 9-9<br>1.21 | 10-0<br>1.31 | 10-4<br>1.43 |
| 2×8 | 16.0″ | 8-7<br>0.24 | 9-4<br>0.31 | 10-1<br>0.39 | 10-10<br>0.48 | 11-6<br>0.57 | 12-1<br>0.67 | 12-8<br>0.77 | 13-3<br>0.88 | 13-9<br>0.99 | 14-4<br>1.10 | 14-10<br>1.22 | 15-3<br>1.35 | 15-9<br>1.48 | 16-3<br>1.61 | 16-8<br>1.75 |
| | 24.0″ | 7-0<br>0.19 | 7-8<br>0.25 | 8-3<br>0.32 | 8-10<br>0.39 | 9-4<br>0.46 | 9-10<br>0.54 | 10-4<br>0.63 | 10-10<br>0.72 | 11-3<br>0.81 | 11-8<br>0.90 | 12-1<br>1.00 | 12-6<br>1.10 | 12-10<br>1.21 | 13-3<br>1.31 | 13-7<br>1.43 |
| 2×10 | 16.0″ | 10-11<br>0.24 | 11-11<br>0.31 | 12-11<br>0.39 | 13-9<br>0.48 | 14-8<br>0.57 | 15-5<br>0.67 | 16-2<br>0.77 | 16-11<br>0.88 | 17-7<br>0.99 | 18-3<br>1.10 | 18-11<br>1.22 | 19-6<br>1.35 | 20-1<br>1.48 | 20-8<br>1.61 | 21-3<br>1.75 |
| | 24.0″ | 8-11<br>0.19 | 9-9<br>0.25 | 10-6<br>0.32 | 11-3<br>0.39 | 11-11<br>0.46 | 12-7<br>0.54 | 13-2<br>0.63 | 13-9<br>0.72 | 14-4<br>0.81 | 14-11<br>0.90 | 15-5<br>1.00 | 15-11<br>1.10 | 16-5<br>1.21 | 16-11<br>1.31 | 17-4<br>1.43 |
| 2×12 | 16.0″ | 13-3<br>0.24 | 14-6<br>0.31 | 15-8<br>0.39 | 16-9<br>0.48 | 17-9<br>0.57 | 18-9<br>0.67 | 19-8<br>0.77 | 20-6<br>0.88 | 21-5<br>0.99 | 22-2<br>1.10 | 23-0<br>1.22 | 23-9<br>1.35 | 24-5<br>1.48 | 25-2<br>1.61 | 25-10<br>1.75 |
| | 24.0″ | 10-10<br>0.19 | 11-10<br>0.25 | 12-10<br>0.32 | 13-8<br>0.39 | 14-6<br>0.46 | 15-4<br>0.54 | 16-1<br>0.63 | 16-9<br>0.72 | 17-5<br>0.81 | 18-1<br>0.90 | 18-9<br>1.00 | 19-4<br>1.10 | 20-0<br>1.21 | 20-6<br>1.31 | 21-1<br>1.43 |

**CHART 5 – HIGH SLOPE RAFTERS** — Allowable spans for slope over 3 in 12; 30 lbs. per sq. ft. live load. (Heavy Roof Covering)

| RAFTER SIZE | RAFTER SPACING | 500 | 600 | 700 | 800 | 900 | 1000 | 1100 | 1200 | 1300 | 1400 | 1500 | 1600 | 1700 | 1800 | 1900 |
|---|---|---|---|---|---|---|---|---|---|---|---|---|---|---|---|---|
| 2×4 | 16.0″ | 4-1<br>0.18 | 4-6<br>0.23 | 4-11<br>0.29 | 5-3<br>0.36 | 5-6<br>0.43 | 5-10<br>0.50 | 6-1<br>0.58 | 6-5<br>0.66 | 6-8<br>0.74 | 6-11<br>0.83 | 7-2<br>0.92 | 7-5<br>1.01 | 7-7<br>11.1 | 7-10<br>1.21 | 8-0<br>1.31 |
| | 24.0″ | 3-4<br>0.14 | 3-8<br>0.19 | 4-0<br>0.24 | 4-3<br>0.29 | 4-6<br>0.35 | 4-9<br>0.41 | 5-0<br>0.47 | 5-3<br>0.54 | 5-5<br>0.61 | 5-8<br>0.68 | 5-10<br>0.75 | 6-0<br>0.83 | 6-3<br>0.90 | 6-5<br>0.99 | 6-7<br>1.07 |
| 2×6 | 16.0″ | 6-6<br>0.18 | 7-1<br>0.23 | 7-8<br>0.29 | 8-2<br>0.36 | 8-8<br>0.43 | 9-2<br>0.50 | 9-7<br>0.58 | 10-0<br>0.66 | 10-5<br>0.74 | 10-10<br>0.83 | 11-3<br>0.92 | 11-7<br>1.01 | 11-11<br>1.11 | 12-4<br>1.21 | 12-8<br>1.31 |
| | 24.0″ | 5-4<br>0.14 | 5-10<br>0.19 | 6-3<br>0.24 | 6-8<br>0.29 | 7-1<br>0.35 | 7-6<br>0.41 | 7-10<br>0.47 | 8-2<br>0.54 | 8-6<br>0.61 | 8-10<br>0.68 | 9-2<br>0.75 | 9-6<br>0.83 | 9-9<br>0.90 | 10-0<br>0.99 | 10-4<br>1.07 |
| 2×8 | 16.0″ | 8-7<br>0.18 | 9-4<br>0.23 | 10-1<br>0.29 | 10-10<br>0.36 | 11-6<br>0.43 | 12-1<br>0.50 | 12-8<br>0.58 | 13-3<br>0.66 | 13-9<br>0.74 | 14-4<br>0.83 | 14-10<br>0.92 | 15-3<br>1.01 | 15-9<br>1.11 | 16-3<br>1.21 | 16-8<br>1.31 |
| | 24.0″ | 7-0<br>0.14 | 7-8<br>0.19 | 8-3<br>0.24 | 8-10<br>0.29 | 9-4<br>0.35 | 9-10<br>0.41 | 10-4<br>0.47 | 10-10<br>0.54 | 11-3<br>0.61 | 11-8<br>0.68 | 12-1<br>0.75 | 12-6<br>0.83 | 12-10<br>0.90 | 13-3<br>0.99 | 13-7<br>1.07 |
| 2×10 | 16.0″ | 10-11<br>0.18 | 11-11<br>0.23 | 12-11<br>0.29 | 13-9<br>0.36 | 14-8<br>0.43 | 15-5<br>0.50 | 16-2<br>0.58 | 16-11<br>0.66 | 17-7<br>0.74 | 18-3<br>0.83 | 18-11<br>0.92 | 19-6<br>1.01 | 20-1<br>1.11 | 20-8<br>1.21 | 21-3<br>1.31 |
| | 24.0″ | 8-11<br>0.14 | 9-9<br>0.19 | 10-6<br>0.24 | 11-3<br>0.29 | 11-11<br>0.35 | 12-7<br>0.41 | 13-2<br>0.47 | 13-9<br>0.54 | 14-4<br>0.61 | 14-11<br>0.68 | 15-5<br>0.75 | 15-11<br>0.83 | 16-5<br>0.90 | 16-11<br>0.99 | 17-4<br>1.07 |

* Reproduced from the 1979 edition of *Dwelling Construction Under the Uniform Building Code*, copyright 1979, with permission of the publisher, the International Conference of Building Officials.

# INDEX